SpringerBriefs in Climate Studies

More information about this series at http://www.springer.com/series/11581

Michael Gordy

Disaster Risk Reduction and the Global System

Ruminations on a Way Forward

 Springer

Michael Gordy
Ornex
France

ISSN 2213-784X   ISSN 2213-7858 (electronic)
SpringerBriefs in Climate Studies
ISBN 978-3-319-41666-3   ISBN 978-3-319-41667-0 (eBook)
DOI 10.1007/978-3-319-41667-0

Library of Congress Control Number: 2016944352

Printed on acid-free paper

This Springer imprint is published by Springer Nature
The registered company is Springer International Publishing AG Switzerland

Contents

Chapter 1
Introduction

Disasters focus the mind, whether they are natural, man-made, or some combination of the two. With the burgeoning of disasters of all kinds, it is no surprise that reducing the risk of their occurrence and preparing responses to them has moved to the forefront of human concerns. Nowhere is this effort more apparent than in United Nations organizations whose mandates include disaster risk reduction (DRR), and whose work has brought together the most substantive and far-reaching information on the subject.

Over the past few years I have worked with several of these organizations. One of them, the United Nations International Strategy for Disaster Risk Reduction (UNISDR), publishes a biennial report, the Global Assessment Report (GAR), which aggregates and analyzes data on DRR efforts throughout the world.

These reports, which can be accessed through links found in the reference section of this essay, have appeared four times so far—in 2009, 2011, 2013, and 2015. The most recent edition was presented at the Third UN World Congress on Disaster Risk at Sendai, Japan, in March 2015 and contributed significantly to formulating the Sendai Framework for Disaster Risk Reduction 2015–2030.

I have made an extensive study of the first three editions of the GAR, which cover almost 700 pages. In this essay I present what I believe to be their most important points. Based on my study I believe that these points, taken together, issue in a more fundamental insight, one that needs to be understood and integrated into all aspects of DRR work. In short, I have taken what I believe to be the conclusions implicit in these three reports and have made them explicit.

The purpose of this essay is to present that distillation of important points organized around this insight. In addition I present a few principles that I believe can make genuinely effective action possible. This essay will, I hope, help anyone who is interested in DRR, either professionally or non-professionally, to work his or

© The Author(s) 2016
M. Gordy, *Disaster Risk Reduction and the Global System*,
SpringerBriefs in Climate Studies, DOI 10.1007/978-3-319-41667-0_1

her own way through the reports themselves. I also hope that it will inspire a reconceptualization of the nature of disasters.

While the factual information in this document is derived from the first three GARs, the interpretation of these facts is entirely my own. I have benefitted from extensive conversations with the editors of the 2015 GAR while it was being put together, but nothing in this essay should be attributed to any of the organizations for which I have worked. The viewpoints expressed are mine.

This is a thought piece, not a research paper, and it reflects the meaning I have found in what I have studied. Readers interested in the specific facts underlying this essay can find them in the reports.

Each of the GARs describes and evaluates the efforts being made around the world to reduce the risk of disasters of all kinds, warn of their onset, reduce their impacts on lives and livelihoods, and strengthen preparedness at all geographic levels. They are quite detailed, with many explanatory examples and much aggregated data. As such they require a significant investment of time and intellectual effort to read and digest. This is as it should be, given that the topic is so important and complex, yet policy makers often do not have the option of reading them thoroughly, due to the many other demands on their time and resources. That said, they express considerable interest in the topic, not least because the incidence of disasters is increasing and is affecting many of the activities for which they exercise oversight.

In response to this conundrum, a summary of each GAR was written and circulated after the report was published. Every summary concerned one specific report and simply compiled its main points. As such, each served as a useful tool for giving policy makers the gist of what had been written and made it easier for them to find elaboration of specific points in the full report.

The present essay will do something different. Its aim is to consider the first three GARs as parts of an ongoing process and to look at the evolution and interaction of the topics they address. One reason for doing this is that, over the period of time covered in the three reports, new information and additional insights have emerged that in some cases have altered the perception of the importance of the issues considered and have brought to light new priorities for DRR. Some understandings that were only implicit when the first report was written have become predominant, while others that seemed to be primary concerns at the time they were expressed have taken on a supporting role. The aim of this examination is to render explicit the results of this evolutionary dynamic.

My purpose is to look at DRR from a non-traditional perspective. Rather than seeing disasters as mainly exogenous, i.e., external to human activity and something from which 'our way of life' needs to be protected, the approach I am taking recognizes that, on the contrary, disasters are primarily endogenous, part and parcel of that way of life, and generated largely by the manner in which we produce, distribute and consume. The analysis includes specific reference to the way power in the world is distributed and exercised (the fundamental feature of governance), and the principles proposed at the end take these power relations explicitly into account.

I also pay more attention than is usual to extensive risk, i.e., the risk of smaller-scale disasters that tend to recur and that extend over wide areas, disproportionately affecting the daily lives of poorer populations. Intensive disasters such as volcanic eruptions are initially more destructive than extensive ones and they do not regularly recur; their deleterious effects are more immediate than cumulative, although those effects may be quite long lasting. They are also much more visible than extensive disasters because they tend to affect all classes in society, albeit unequally. For those (relatively few) with access to disaster insurance, their financial impact can be mitigated. Extensive disasters tend to be more or less invisible to everyone except those directly affected by them, and they have received virtually no attention in DRR planning until fairly recently. The discussion here is intended to help bring extensive risk more fully to light.

This essay is interpretive and does not cover all the issues addressed in the original reports. Its perspective sees history as retrospectively deductive but prospectively inductive, entailing what the Danish philosopher Soren Kierkegaard described as the tragedy of the human condition, namely, that life has to be lived forward even though it is only understood backward. It is thus a 'backward' glance at the evolution of the first three GARs, undertaken in hopes of obtaining clearer intimations of what sorts of activities DRR should undertake in the future.

Any historical understanding is necessarily selective. It is not possible to recapture the 'innocence' of contemporaneous perspectives because we see the past through the lens of our subsequent experience and reflection. Over time, some things will take on greater importance than others according to our subjective (and by this is meant 'collectively subjective') point of view. Therefore, this examination is not presented as definitive in any way. On the contrary, its purpose is to contribute to a much wider and more profound discussion that should extend well beyond what has come to be known as the 'disaster risk reduction sector'. It is meant to open up new avenues and orientations of conversation and investigation, not to terminate reflection on any of the topics considered.

The underlying theme of the endogenous nature of disasters runs through everything, but it is approached from a dozen different angles, each based on issues that are found in the three GARs. These are:

    I. Underlying Risk Factors
   II. The Social Construction of Risk
  III. Development and Risk
   IV. The Poverty-Disaster Risk Nexus
    V. Global Capitalism and Disaster Risk
   VI. Political Will and Governance
  VII. The Importance of Extensive Risk
 VIII. The Invisibility of Risks
   IX. Urbanization and Land Use
    X. Agribusiness and Food Security
   XI. Environmental Degradation
  XII. Climate Change as a Magnifier and Meta-disaster

These twelve topics are ordered in a virtuous circle, beginning with overall considerations that are necessarily general, moving through more specific topics, and ending up with an issue that synthesizes the general with the specific. In this way, what was somewhat abstract in the beginning will become more concrete by the end.

There is a particular logic to this approach. The initial four sections describe certain characteristics of disaster risk as they first appear, i.e., separately and with only a few intimations of what binds them together. The next two sections outline briefly some features of that binding structure, while the rest of the essay revisits parts of the initial descriptions on the basis of a more explicit understanding of their dynamic interconnection. An appropriate metaphor might be climbing a spiral staircase while looking back at the entire itinerary of the steps one has taken.

After completing this trajectory, the document offers some principles for practical action based on the resulting perspective.

Chapter 2
Underlying Risk Factors

The Fourth Priority of the Hyogo Framework for Action (HFA) mandates a reduction in the underlying factors of disaster risk. Honoured too often in the breach, this particular mandate is nonetheless the heart of the HFA and has become increasingly important throughout the period covered by the three published GARs. Global surveys of people involved with DRR have shown that increasing the emphasis on reducing these risk factors is an almost universal concern, one that is intensifying all the time.

The risk factors most often cited include vulnerable rural livelihoods, poor urban and local governance, ecosystem decline, and climate change, with climate change acting as a kind of 'meta-factor' inasmuch as it magnifies the others (we will look at this more closely in a later section). Others mentioned are rapid and unplanned urbanization, the expansion of agribusiness (which is often associated with increasing urbanization and environmental degradation) and, most importantly, poverty.

Poverty reduces the ability of people to protect themselves from hazards, obviously because the resources that would be necessary for, say, constructing disaster-resistant lodgings are simply not available to them. In addition, when a disaster strikes, poor people often do not have any assets to buffer their livelihoods or to help them recover from economic losses. Poverty thus exacerbates the effects of disasters, while disasters increase poverty. It is a vicious circle that is referred to in the GARs as the Disaster Risk (DR)-Poverty Nexus.

© The Author(s) 2016
M. Gordy, *Disaster Risk Reduction and the Global System*,
SpringerBriefs in Climate Studies, DOI 10.1007/978-3-319-41667-0_2

2.1 Rural Poverty

Poor rural areas, especially in isolated or remote regions, are often subject to highly vulnerable housing as well as to weak or non-existent emergency and health services and infrastructure. Likewise, poor rural communities have limited access to productive assets such as land, fertilizers, irrigation, and financial services. All this is associated with political marginalization as well as with discrimination and exclusion of various kinds due to race, gender, or ethnicity, creating structural vulnerabilities that increase with each disaster. If the rural population is far from an urban centre, its markets tend to be considerably weaker than those in communities closer to the more steady and concentrated exchange of commodities characteristic of cities and their surroundings. And underlying all of this is the fact that rural livelihoods are subject to the vicissitudes of national and global markets for agricultural products. We will examine this aspect in more detail later on.

Agricultural livelihoods are exposed directly to weather-related hazards and are most vulnerable to unrecoverable losses from meteorological disasters. Localized hazards such as storms, frosts, heat waves, cold spells, and minor droughts can mean the loss of an entire harvest, while major disasters such as serious, long-term droughts can destroy agricultural production of plants and animals over wide areas for several years. Because the rural poor tend to have access only to the least-productive land, their vulnerability to disaster is greatest and their resilience least. They also suffer an incapacity to recover from crop or livestock losses, which means they are easily pushed to destitution by a single disaster. In regions where there is no significant government safety net, which is the case for most if not all poor countries, there is little to brake their fall into the disaster-poverty spiral.

Even if a particular disaster does not push people into this spiral, over time repeated disasters ratchet up poverty by increasingly weakening livelihoods and steadily undermining the ability of populations to recover, pushing rural households further into chronic poverty and deprivation. This point is particularly relevant to extensive risks, a subject that will be taken up later.

On the other hand, while disasters increase poverty, poverty is an important ingredient in turning hazards into disasters. For example, if poor farmers have access only to relatively unproductive land, they will be forced to overuse the land they have by over-grazing, deforestation, and unsustainable extraction of water resources that magnify hazard levels and aggravate disaster risk. But when people are pushed to their limits, they must focus on surviving in the present even though the strategies for doing so may only exacerbate their problems in the longer term. This exemplifies the business principle that long-term assets do not relieve short-term debts, and is an example of a tactic that might make sense at the micro-level but which almost invariably turns out to be counterproductive or even self-destructive over the longer term.

With food production becoming increasingly globalized at the hands of agribusiness, along with speculative fluctuations in the world prices for agricultural products, small farms have become less viable and the rural poor have been forced to find either supplemental or replacement livelihoods in non-farm occupations, including food processing, transport, manufacturing, and even finance. In addition, many farmer workers migrate to the cities, either seasonally or permanently, in order to survive and provide remittances to household members who remain in the countryside. While some government aid programs and the efforts of NGOs have made it possible for a relative few of these farmers to remain on the land, the net result has been a burgeoning influx of migrants to the cities, with the attendant problems associated with unplanned and uncontrolled urbanization.

2.2 Urbanization and Poverty

Over half of the world's population lives in urban areas, and this proportion is expected to rise to 70 % in the next few decades. Almost three-quarters of these people live in low- and middle-income countries, with a total of over 900 million of them extremely poor and lacking protection from common life- and health-threatening diseases and injuries. It is expected that almost all of the world's population growth between now and 2025 will take place in urban areas in these poorer regions.

The urban poor are subject to increased disaster risk because of two inter-related processes. First, outward urban and economic development generate new patterns of extensive risk such as flooding and other weather-related hazards, particularly affecting informal settlements on the periphery of large cities as well as in small and medium-sized urban centres. Second, as cities grow they become more densely populated, and there is an inward concentration or intensification of disaster risk associated mainly with earthquakes, tropical cyclones, and floods, causing major asset loss and mortality amongst the urban poor. Contending with these two dynamics is a matter of urban and local governance, and much too often this governance is either desperately inadequate or lacking altogether.

The concentration of private capital and its associated economic opportunities are crucial drivers of urban infrastructure expansion, while population increase is fuelled in large part by the rural-urban migration noted above. Concentrated private capital, however, does not by itself ensure that the supply of land for housing, infrastructure, and services keeps up with population growth, nor does it produce the regulatory framework to ensure that the environmental, occupational, and nat-ural hazard-related risks generated by urban growth are managed adequately or at all. In poor countries, there is often a mismatch between the economic drivers of urban expansion and the institutional mechanisms to manage or govern the direct and indirect implications of economic concentration.

As a consequence, urban expansion in poor countries (and elsewhere) often occurs outside the legal framework of building codes and land use regulations. It regularly takes place without officially recorded or legally sanctioned land trans-actions. Inevitably, those with the least purchasing power and the least political influence end up occupying land or housing that nobody else wants. Informal settlements spring up in these areas almost naturally because private investors are not interested in these places for commercial development, while city governments are usually incapable of using them to provide for the housing needs of the urban poor.

These informal settlements are rife with the extensive risks of local flooding, fires, and landslides. In most cities this is due to a significant proportion of informal settlements being built on dangerous sites that lack infrastructure and services. Because most informal settlements are illegal, they usually have no way to access these necessities. In addition, this kind of urban development magnifies hazard levels. Building on green areas, for instance, often occurs with no provision for more effective drainage and produces water runoffs that create floods, while encroaching construction destroys natural drainage channels or flood plains that would ordinarily help dissipate these floods.

At the level of individual households, the absence of land titles in these informal settlements means that the inhabitants have no incentive to improve the standard of their housing, nor do they have access to housing finance or technical assistance. This multiplies the effects of extensive risk, because events that are merely dis- comfiting in more well-off settlements become daily disasters when they affect defenseless informal ones. Additionally, because there are no land titles and hence no legal record of these settlements, governments often ignore them when it comes to providing infrastructure and services, which intensifies the negative impact of disasters among the poor.

Conditions are worse for poor women, both within these informal settlements and outside them, for they are discriminated against with regard to land tenure and access to income and services. These inequalities exacerbate their vulnerability to disaster risk and compound the effects of poverty on their lives. Unfortunately, there is little gender-disaggregated data available for risk assessments, in large part because this has not been a priority for many governments. It is a priority for many donor organizations, however, so there is hope that this situation can be ameliorated somewhat by their pressure on recipient public institutions.

2.3 Bad Governance

Bad urban planning exacerbates these tendencies. The driving force of economic development, fuelled by private capital, skews decisions about where and what to build, leading urban areas to expand into hazardous locations that are sometimes extremely dangerous. Building power plants on fault lines, constructing housing on flood plains, cutting down shoreline mangroves to clear land for various com- mercial purposes, are just a few examples of decisions that put commercial gain ahead of concern for the lives and livelihoods of local populations or even concern for the sustainability of the commercial projects themselves. Without the moder- ating force of effective political regulation of economic development, there is no significant way to incentivize putting DRR concerns on anything like an equal footing with economic gain when development decisions are made. If governance means exercising political authority for the welfare of the community as a whole, then allowing economic development to proceed unfettered is poor governance indeed.

Governments in poor countries aren't stupid, or at least they are no more stupid than governments in rich countries. Poor governance is more a result of much deeper structural irrationalities than it is of political backwardness, whatever 'backwardness' might mean. Earlier it was noted that the way we produce, distribute, and consume, i.e., the way our economic life is structured, is the primary generator of disaster risk. Specifically, however, it is important to note that economic development does not, and in fact cannot occur without some fundamental relationship with political power. The underpinnings of economic life are guaranteed politically, while politics in the world as we know it tends to reflect the distribution of economic power in society. Governments and governance necessarily reflect this essential inter-relationship.

This is why it is important to look at disasters from the point of view of political economy rather than from the perspective of economics as separate or separable from politics. That said, it follows that the nature of economic development is a governance issue of the first order. It reflects the specific relationship between economic and political power in every country and in the world as a whole. Placing DRR at the heart of development decisions will require a fairly important shift in that relationship and in the way we think about the obstacles that stand in the way of reducing the underlying factors of disaster risk. It will require changing the balance of forces between economic and political power so that the latter reflects the interests of the whole community rather than being the handmaiden of commercial developers. While this is not the place to investigate what this means in practice (in part because there can be no blueprint), it is useful to keep it in mind when assessing DRR progress and planning transformative actions to diminish the underlying factors of disaster risk.

As an aside, we should remind ourselves that bad governance is caused in part by bad information. Decisions cannot be usefully made if the information on which they are based is partial, skewed, or even false. But what counts as 'information'? In too many cases, determining economic losses depends entirely on monetary values and thereby underestimates the real costs of disasters to the poor. In rich countries, the monetary value of losses may be high, but their effect on the livelihoods of the affected population may be nowhere near as devastating as the effects of much lower monetary losses on poor people.

What are the real economic and humanitarian losses to a poor family or community that has had its economic infrastructure totally destroyed, even though the monetary value of that infrastructure was low? If they do not have the means to rebuild, their future is in serious jeopardy. If, for example, a school in a poor community is wiped out by a cyclone, what does this mean for the futures of the children and young people who are robbed of their chance for an education? How do you put a price on that? It is not the assessed financial cost of rebuilding the school that is at stake here, but any number of valuable assets that escape monetization. Yet so many decisions taken by political authorities accept business accounting (i.e., the monetization of everything) as the only legitimate framework for calculating disaster risk. When looking through this lens, it is no wonder that so much remains invisible.

2.4 Ecosystem Decline

Ecosystems provide the fundamental necessities of life, such as food, water, protection from extreme weather events, and the purity of the air we breathe. These systems, however, are threatened by unregulated economic expansion and are fast losing their capacity to moderate disaster risk. We are polluting the atmosphere to such an extent that the planet's counterbalancing natural waste removal mechanisms, the so-called 'carbon sinks', can no longer keep up, with the buildup of atmospheric toxins rendering a fast-growing number of areas virtually unlivable. We are using (and abusing) water supplies at a rate that is completely unsustainable in the medium- and long-term, and this is already impacting human life negatively in the near term as well. Unbridled urbanization and uncontrolled agricultural and extractive practices are cutting into the sustainability of food production in the interests of short-term profitability, while small-holding farmers are rapidly being squeezed off the land, with those remaining forced to engage in short-sighted practices such as over-grazing simply in order to survive. Add to this the intensifying volatility and unpredictability of commodity prices due to financial speculation, and you have a recipe for a concatenation of extensive disasters, with intensive disasters as their cumulative outcome.

Modifications of ecosystems to increase production of food and fiber have unintentionally led to increasing people's exposure to risk, such as when deforestation of hillsides for agricultural purposes increases the danger of landslides and diminishes protection from high winds and flash flooding. Likewise, as part of deforestation worldwide, destruction of mangroves is removing barriers against storm surges. Since forests play a crucial role in protecting and regulating soil and water catchments, deforestation, which is increasing globally at a rate of 0.2 % a year, is contributing to intensified flood and drought cycles. Many ecosystem modifications like this have been accomplished in the service of economic interests that remain unaffected by the associated hazards that accrue to local, often poor, populations.

Land degradation affects almost 15 % of the global population, primarily in poor areas, although the figure is probably larger if we include areas affected by bad agricultural practices in rich countries as well, where soil exhaustion and erosion, overuse of agricultural chemicals, land poisoning by the extractive industries, and other short-sighted, unsustainable practices are ubiquitous. By poisoning water reserves while at the same time overusing them, these practices are destroying supplies of nutritious food and clean water that are essential to the continuation of human life. Yet these phenomena are overshadowed as underlying disaster risk drivers by the ongoing, man-made menace of global climate change.

2.5 Climate Change

Climate change is denied only by those with a strong economic interest in doing so, and its anthropogenesis is discounted for the same reason by those who accept its reality but do not want to try to do anything about it. A virtual unanimity of climate scientists, however, agrees that climate change is real, increasingly dangerous to human life, and man-made.

Climate change affects nearly every aspect of human activity, and generally does so destructively. With respect to disaster risk, its effects are devastating and are becoming more so. While climate change does not magnify all disasters, it surely intensifies the 80 per cent of disasters that are weather-related. And even though a great deal can be done to diminish disaster risk through actions that are not directed at climate change, it seems clear that in the coming years climate-related disasters, both intensive and extensive, will demand significantly increased attention to this threat to the continuation of human life.

It may even be misleading to list climate change alongside the other underlying risk factors we have been considering. It probably deserves its own category as a *meta-disaster*. By its very nature it does not occur locally, even though local conditions shape its impact in the moment. It does not submit to local actions except where they consist of adaptations to its effects; mitigating or stopping climate change will require global policies that are universally enforced and adhered to under threat of mandatory sanctions.

This makes climate change different from poverty alleviation, unchecked urbanization, or ecosystem protection because these can be dealt with at the level of communities, countries, and regions. It is possible to rebuild informal settlements, improve governance at all geographic levels, repair environmental damage to some extent, and create resilient communities to withstand various specific disasters. Building informal settlements in flood-prone areas can be stopped, although that would probably involve a fairly massive redistribution of resources, not to mention political and economic power, and the disaster-poverty nexus can certainly be broken by social policies qualitatively and quantitatively different from the ones currently in place. We cannot, therefore, put off addressing the structural causes of poverty, upgrading housing in informal settlements, strengthening governance at all levels and managing natural resources much better in low- and middle-income countries (with attention to these issues in rich countries as well). Climate change cannot be used as an excuse to relax our efforts to deal directly with these and other factors underlying disaster risk.

But climate change rolls on, generating more intense weather-related events and in so doing increasing poverty and sapping people's resilience. As a set of countermeasures, what is required is a complete shift in the global economy away from burning fossil fuels, away from overdependence on carbon-based agriculture, towards a reorganization of global transport, and finally towards a complete overhaul of the way we order our lives, with the consequential shift in political power that all this entails.

This may seem daunting, and of course it is, but what other choice do we have? Technological fixes belong to the realm of fantasy, based on a quasi-religious belief in humanity's capacity for 'innovation' (always seen as technological) and a deep-seated desire to change, "but not too much". Even just mentioning the implications and requirements of climate change takes most of us out of our "comfort zones", (an infelicitous locution if there ever was one, and a polite way of telling people to shut up). There are, however, ways of approaching climate change that identify the lever with which the problem can be made to budge. We will come back to this toward the end of this document.

Chapter 3
The Social Construction of Risk

Now that we have looked briefly at the underlying risk factors for disasters and have set out the overall terrain of our discussion, it is time to take a closer look at specific topics so that, ultimately, we can see how they interact with and support one another. The first topic is the social construction of disaster risk.

Analyzing disaster risks has become increasingly difficult over recent decades because of the growing interconnectedness and interdependency of modern societies. Exposure to hazards has always existed, but the effects of those hazards have become more complex and new vulnerabilities have emerged, with the result that hazardous events turn more readily into disasters. It is the magnitude of losses from natural events that turns them into disasters, and this magnitude is significantly dependent on the aggregated actions of human beings, past and present. As the central theme of this document becomes increasingly clear, we will see that these actions have a dynamic structure that ramifies their effects beyond what can reasonably be attributed to human error or cupidity.

There is a growing probability of simultaneous crises, where different hazardous events occur at the same time; sequential crises, where hazardous events trigger cascading disasters; and synchronous failures, where these different crises converge. An example of the latter is the earthquake, tsunami, and nuclear meltdown configuration in Japan in March, 2011, where a large earthquake triggered a tsunami, and the combined effects created the conditions for a disaster at a multi-reactor nuclear power plant at Fukushima. The earthquake disrupted Japan's power grid, which was then unable to supply the electricity needed to cool the nuclear plant's reactors, while the tsunami disabled the plant's backup generators, leading to the worst nuclear meltdown since the Chernobyl disaster of 1986, the effects of which are still ongoing and severe.

An example of a sequential crisis is the heat wave and its associated wildfires that hit western Russia and Ukraine in 2010. There, severe drought created the

© The Author(s) 2016
M. Gordy, *Disaster Risk Reduction and the Global System*,
SpringerBriefs in Climate Studies, DOI 10.1007/978-3-319-41667-0_3

conditions for wildfires, which exposed vulnerabilities that cascaded into disastrous impacts in areas as diverse as public health and air traffic.

These examples are important because they demonstrate concretely what is meant by the social construction of risk. In each case social and economic practices, ultimately facilitated by governance decisions, created the context for the concatenation of hazards and vulnerabilities that made these disasters possible.

The Fukushima disaster in Japan happened in part because a nuclear power plant was built on a very dangerous fault line, part of the so-called "ring of fire" that follows the Pacific Rim. That decision, ultimately approved by the Japanese government, was based on assurances by a large and reputable multinational corporation, General Electric, that they could build a nuclear power plant capable of withstanding the most severe shocks ever recorded in the area, and that their backup systems would function adequately in any case.

It is not known whether plant specifications reflected an absolute commitment to safety or whether cost-cutting and/or profit enhancing considerations intervened along the way, but it is clear that the decision to build at this location was more risky than wise, given the inevitable uncertainty about the intensity of future earthquakes and the possibility that proximity to the ocean might make the plant vulnerable to a tsunami as well. The reason that this decision was taken was undoubtedly governed to a large extent by energy needs and cost effectiveness, both of which reflect a certain kind of development paradigm based on quantitative growth and consumerism.

It is clear, however, that there could not have been a disaster of the proportions of Fukushima, with its attendant large-scale food and water shortages and widespread radiation dangers, if a nuclear plant had not been built on that spot. A wealthy country like Japan can protect its people by building resilient communities that can withstand earthquakes and tsunamis to the extent humanly possible, but it cannot protect them from the effects of a nuclear meltdown once one has occurred, nor can it prevent such meltdowns by building nuclear plants on fault lines. This was a case of bad governance based on a dominant economic paradigm, revealing the interaction of economics and politics in generating disaster risk.

The heat wave in Russia and Ukraine and its associated wildfires also needs to be understood in light of the social construction of drought risk. There appear to be a number of drivers, all socially derived, that translate water shortage over time (meteorological drought) into disaster risk. These include rural poverty; increasing water demand due to urbanization, industrialization, and the growth of agribusiness; poor soil and water management; weak or ineffective governance; and climate change.

There are three categories of drought: meteorological, agricultural, and hydrological. Meteorological drought refers to a precipitation deficit over time, while agricultural drought refers to conditions where this deficit renders the soil incapable of supporting crops, pastures, and rangelands. Hydrological drought occurs when the water in lakes, rivers, reservoirs, streams, and groundwater diminishes to the extent that non-agricultural activities are impacted. These activities include, among others: tourism, recreation, urban water consumption, energy production, and ecosystem conservation.

All three types of drought have been affected over time by human activity, particularly economic development. For instance, annual global water consumption tripled between 1960 and 2011, while the global population little more than doubled. The difference in proportion here can be ascribed to the explosion of economic growth during that period and especially to the unequal, wasteful consumption patterns associated with the way the benefits of growth have been distributed.

This increased water use has in many cases transformed meteorological drought into the agricultural and hydrological varieties, with grave human consequences. In addition, changes in wind and other weather patterns connected with climate change have altered precipitation patterns in many parts of the world, engendering meteorological drought where such things have heretofore rarely occurred. Since climate change is accepted overwhelmingly by climate scientists to be the result of human activity, we can safely say that more frequent and widespread meteorological drought is generated, at least indirectly, by what we do and the way we do it.

In the case of the heat wave and wildfires in Russia and Ukraine, long-term disinvestment in forest maintenance following the dissolution of the Soviet Union, along with (sometimes speculative) construction in hazardous zones, contributed heavily to the cascade of disasters initiated by drought and other climatic factors. Inattention to preserving water reserves, coupled with expanded water consumption and diminished forest maintenance, exacerbated the effects of meteorological conditions, turning the drought hydrological and setting the conditions for the wildfires.

To address the underlying drivers of drought risk, countries will have to reorient their risk governance, especially when it is related to development planning as well as to land use and water management. There are powerful disincentives to doing this due to the political implications of property rights, but ever-increasing drought risk may soon outweigh them. Risk that is socially caused can often be socially alleviated.

These two examples are among many that can be given, such as building on flood plains, short-sighted agricultural practices that include poisoning and exhausting the soil, shoddy building construction, and the aforementioned deforestation, etc. All are instances of social risk construction, because each in its own way intensifies hazardous situations and enhances their negative consequences. All reflect, in one way or another, underlying social practices and norms. Since the social practice *par excellence* is economic practice, the subject to which we shall now turn is the relationship between economic development and disaster risk.

Chapter 4
Development and Disaster Risk

The importance of the effects on disaster risk of the dominant development paradigm is a topic that gained importance throughout the period covered by the three published GARs. The first report, while paying close attention to the poverty-DR nexus, did not explicitly single out economic development as a whole for examination, simply noting that many of the institutional and legislative structures created for DRR have had little influence on poverty reduction, particularly in countries where much development takes place informally and is unregulated. It did note, however, that while economic development can reduce vulnerability by increasing the resources for creating resilience, it exacerbates the exposure of people and assets in areas prone to hazards. This comment served as a precursor to raising thinking to a level where the disaster-generating characteristics of the dominant economic development paradigm are put front and centre, which is what happened in later reports.

The second GAR, from 2011, moved this evolution along a bit by indicating some of the ways that economic growth raises risk levels. It noted the difficulty governments have in finding economic and political incentives just to identify the costs and benefits of having a balanced portfolio of disaster risk management strategies. Part of the problem, it implied, lies in having to think about these strategies confined to the language and metrics of business accounting. Since not all risks can be monetized, it is difficult to make a convincing argument for DRR strategies in financial terms, especially if those strategies limit capital accumulation in any way. Yet moral and humanitarian arguments tend to carry little weight in these debates.

In any event, it was pointed out that the expansion of risk due to rapid economic growth during the previous decade has outstripped the ability of governments to address these risks adequately. This has been particularly true for low- and middle-income countries. As mentioned above, decreases in vulnerability are being

© The Author(s) 2016
M. Gordy, *Disaster Risk Reduction and the Global System*,
SpringerBriefs in Climate Studies, DOI 10.1007/978-3-319-41667-0_4

overwhelmed by the increasing exposure of populations and economic assets due to economic expansion.

GAR 2013 took up the question more explicitly. That report was aimed at convincing people in the private sector that it makes good business sense to integrate DRR into business strategies and investment decisions. Through this, however, it became clear that the entire economic paradigm of unbridled quantitative monetary growth stands in the way of achieving this integration. Although it was not stated openly, many elements of the report pointed toward the disaster-generating effects of the regime of private accumulation.

By noting the interdependence of the public and private sectors, GAR 2013 reflected the point made earlier in this document about the need to look at social reality from the point of view of political economy. This perspective recognizes that politics and economics are deeply implicated in one another without being reducible to each other. Business cannot function without government assuring the enforcement of a legal infrastructure that secures property rights and labour relations as its top priority, while governments in capitalist societies cannot survive without assuring that business keeps economic life going for their populations. It became apparent in the report that governments come up against incredibly daunting obstacles to imposing DRR remedies on their business communities, not the least of which is that governments are often beholden in many ways to those communities for their political power. Economics trumps politics too frequently, demonstrating that when a former president of General Motors said that "the business of America is business," he was not joking, nor was he overstating his views.

Countries compete to attract investment. This is one specific way that government and business intersect, because governments provide the infrastructure, tax incentives, and financial emoluments to try to win this competition. Yet part of what a government can offer is a safe environment for business to operate in, and this is one powerful reason that DRR is beginning to be taken more seriously around the world, including outside the so-called DRR sector.

Government support for DRR programmes is becoming part of a country's competitive advantage, much like its assurance of political and social stability, its maintenance of low labour costs, its willingness to forego a significant amount of tax revenues from business, its business-friendly legal system, its flexible environmental regulations, its lax controls over profit repatriation, and its provision of a vocationally educated workforce. What distinguishes DRR from other elements in this enticement package is that it benefits the broad mass of the population, at least in principle. It also protects the other elements of the country's competitive advantage from disasters, while safeguarding some of the fundamental conditions that must be fulfilled if businesses are to be able to operate. Nonetheless, when we look at the way businesses act in many unregulated locations, it becomes apparent that winning the competitive advantage competition is a mixed blessing, at least from the point of view of DRR.

A country's wealth and future development possibilities depend to a significant degree on its so-called 'natural capital', i.e., its set of renewable and non-renewable natural resources, including agricultural land, fisheries, fossil fuels, forest resources, water, biodiversity, and minerals. A country with a declining base of natural capital is not likely to achieve a sustainable increase in wealth, nor is it likely to become more resilient in the face of disasters. Yet many of the risks generated through business investments are externalized and transferred through climate change, land degradation, and over-exploitation of water resources. This has deleterious and often irremediable effects on natural capital.

In the long run these externalized risks are shared not only in space, over widening geographic areas, but in time, because the exhaustion of natural capital compromises the livelihoods and health of future generations. In other words, while the economic benefits of these investments are privatized, many important social, environmental, and future economic costs are socialized. The costs of these transferred risks are rarely factored into development decisions, whether by investors or by governments whose mandate it is to serve the well-being of the population as a whole.

For instance, the depletion of natural capital is accelerated by business investments in activities such as mining or gas and oil extraction. It is also accelerated by agribusiness. These investments are associated with the demand for raw materials and energy from rapidly expanding and urbanizing economies whose expansion is also connected with business investments, as well as with the demand to supply food to a growing urban population.

Business investment in bio-fuels, timber, and agribusiness, especially those requiring tropical rainforests to be cleared, increase wild-land fire hazard and lead to major depletions of natural capital, as well as to the loss of critical, shared ecosystem services, thus increasing disaster risk. Agribusiness investments in drought-prone areas similarly contribute to increasing land degradation and to over-exploitation of water resources. Intensive agriculture and over-grazing, often the result of the rural poverty associated with the overall development paradigm, is connected to land degradation as well, due to salinization from inappropriate irrigation, deforestation, and the breakdown of traditional agro-ecological systems. Agricultural land abandoned to urbanization and forests converted to rangeland are comparable indicators of the way investment decisions reflecting the dominant development paradigm lead to the loss of natural capital and thereby increase disaster risk.

Tourism is a business sector where the effects of investment on risk generation can be clearly seen. Particularly in the Small Island Developing States (SIDS), where hazardous beach and waterfront locations abound, tourism has become a mainstay of economic life and is a sector where investment has been, and continues to be, intense. Construction of tourism facilities and infrastructure has burgeoned around the world in many areas where disaster risk is great, but often the magnitude of this risk is hidden from international investors, who are not sufficiently worried about it in any case because of the attractive monetary returns they anticipate.

Moreover, these risks are quite often transferred to households, small enterprises, or the public sector. Coastal erosion from building infrastructure for tourist facilities, such as the large jetty on the beachfront of Mui Ne in Vietnam, is but one example. This construction appears to have caused coastal erosion of the shores of Phan Thiet, the nearby provincial capital, resulting in increased storm surge and flood risk for the local inhabitants. But despite a plethora of these kinds of unwanted and unexpected effects, SIDS and other tourism-dependent countries still compete to attract investment in the tourism sector because that seems to them to be the best way to achieve growth, and in some cases the only way. In so doing, however, they accept, either implicitly or explicitly, part of the disaster risks generated by hotels and resorts. Consequently, the countries that have been most successful in attracting investment in their tourism sectors have increased their hazard exposure and have experienced the highest losses to their GDP and the greatest damages to their uninsured public and private infrastructure.

Overall, in a great many countries there is a trade-off between rapid economic growth and disaster risk reduction, a trade-off reflecting national political priorities. If governments are competing for investment, and if the perceived additional costs of DRR are considered a barrier to obtaining that investment, then many governments, understandably if not justifiably, prioritize growth over risk reduction. In many cases, even though high levels of disaster risk undermine a country's competitiveness, governments have chosen to downplay or ignore this fact, contributing in some cases to investor risk blindness (in other cases, anticipated super-profits are equally blinding).

Later on we will look at some of the ways that the trans-nationalization of capital has decisively exacerbated the competition for investment (for example by entering into free trade agreements and deregulating financial flows), but for now it is sufficient to mention that the trade-off between DRR and the economic growth has an underlying structure that facilitates and reproduces this ultimately self-destructive choice.

The message here is that disaster risk, far from being exogenous to economic practice, is inherent in it inasmuch as this practice is configured according to the dominant paradigm of unbridled quantitative growth and private accumulation. A great many factors illuminating this insight can be found in the three published GARs, particularly in GAR 2013, but a number of them have been present since GAR 2009. One of the most significant of these is the poverty-DR nexus, a phenomenon to which we shall now turn.

Chapter 5
The Poverty-DR Nexus

It should come as no surprise to learn that poorer countries have disproportionately higher risks of mortality and economic loss from disasters than rich ones, given similar levels of hazard exposure. Earlier we listed the underlying disaster risk drivers, namely, rural and urban poverty, urbanization, bad urban governance, and eco-system decline, i.e., environmental degradation, all of which are more present in poorer countries than in rich ones. The unequal distribution of these drivers translates poverty and every day risk into disaster risk. Concomitantly, the poor cannot buffer disaster losses and therefore generally become even poorer when disaster strikes. The weakness of social protection in poor countries (and increasingly in the poorer sections of rich ones) adds to this problem.

Whereas historically the vast proportion of disaster risk has been concentrated in limited areas of the Earth's surface, recent meteorological events associated with climate change have begun to spread hazards to areas previously more or less immune to them. Since over two thirds of the mortality and economic loss from internationally reported disasters is associated with meteorological, climatological, and hydrological extreme events, this fact is highly significant. If we add the largely unreported mortality and losses from extensive risks, the importance of climate-based disaster risk is mind-boggling.

The International Panel on Climate Change (IPCC) has confirmed that changes are already occurring in the amount, intensity, frequency, and type of precipitation across the globe, with geographic distribution expanding apace. This means that more areas are being affected by drought, as well as by the increasing numbers of heavy rainfall events that lead to all types of flooding. In addition, the IPCC notes that the intensity and frequency of certain kinds of tropical storms is growing, along with an expansion of their geographic reach.

Inasmuch as the poor have less power to resist these hazards or to recover from their effects, the expansion and intensification of disaster risk signals an increase in

© The Author(s) 2016
M. Gordy, *Disaster Risk Reduction and the Global System*,
SpringerBriefs in Climate Studies, DOI 10.1007/978-3-319-41667-0_5

their vulnerability and in turn undermines poverty alleviation by making them even poorer. Burgeoning inequality of income and other resources is thus being translated into more disaster risk, while more frequent and intense disasters are accelerating the increase of economic inequality and immiseration.

It would be easy and perhaps even reassuring to ascribe this vicious circle to so-called 'natural' causes, ignoring that climate change is agreed to be man-made by a near-unanimity of climate scientists. It is more comfortable to avert our gaze and continue on with our lives. But there are, in fact, macro-economic and social-structural reasons for the acceleration of inequality, which leaves open the possibility that the problem is subject to human amelioration. Now is not the time to discuss this, but the issue will come up again as we proceed. It resonates with the position implicit in all three GARs that disaster risk is endogenous to the way we order our social and economic life and is thus amenable, if only partially, to being transformed.

The unequal effects of disasters between poor and rich reflect the fact that poor households lose a greater proportion of their assets than the better off. This is almost common sense. If all you own is your uninsured shack and a few tools, and these are swept away by a flash flood, how do you hope to recover without outside help? Whereas if you have assets to fall back on, for example insurance, your prospects are much better; you haven't lost everything. But even if a poor family has a few saleable assets that survive a disaster, the fall in prices due to everyone trying to sell at once compromises the effectiveness of these assets in recovery.

Likewise, distress sales of assets act as poverty accelerators, with people forced to sell productive items just to survive. These goods are then bought up by richer households at undervalued prices, adding to inequality and polarization, particularly in rural areas. Even more distressing, poor farmers suffering a disaster may be forced to sell some of their food production at low prices after a harvest just to meet urgent cash needs, only to buy the food back later in the year at double or triple the original price so that they can bridge the hunger gap until the next harvest.

While food relief, cash transfers, micro-credits, insurance, and public health interventions can help avoid translating disasters into increased poverty, the absence or lack of timeliness of these kinds of help can affect the long-term prospects of highly vulnerable groups such as children. For instance, if children go hungry it can stunt their physical and mental development and make them more vulnerable to disease, thereby ruining their chances of escaping a poverty-stricken life. Likewise, destruction of a school in a community that cannot afford to rebuild it robs children of the chance for an education. These kinds of things cannot be repaired, cannot be monetized, and cannot be compensated for, and they rarely appear in loss estimates. Yet they are real nonetheless and have to be the object of a different manner of evaluating risk.

It is also important to recognize the long-term effects of extensive risk on poverty accumulation. Extensive risks, as we have noted, have to do with smaller, recurring disasters such as flash floods, landslides, drought, wildfires and the like, which extend over fairly wide areas and whose greatest impact is on the most vulnerable. These events, while more manageable than intensive disasters in the first instance, compound their negative effects with repetition and ultimately undermine coping strategies by using up the limited assets available. They also strain morale and lead to a sense of hopelessness and resignation that is very counter-productive.

The cumulative losses from these kinds of events, which tend to shape the daily life of poor communities, eventually add up to those of an intensive disaster. Nevertheless, they do not often attract the attention of governments because they are not insured, nor do the affected populations carry much political weight. Moreover, the accumulated poverty saps resilience and thereby increases disaster risk as well, imprisoning vulnerable communities in another vicious circle.

Poor rural households are particularly likely to be caught in this trap. They are often disproportionately exposed to weather–related hazards and are therefore more vulnerable because their livelihoods are weather-dependent. Historical patterns of land distribution and land tenure, both social constructs, likewise tend to discriminate against the poor, who may only have access to marginal or unproductive land, while the land they do have is often excessively prone to flooding and drought or has poor soil.

These poor households are usually indebted and have little or no surplus capacity to absorb or recover from crop or livestock income losses. A small loss in income may set off a ratchet effect that feeds into further poverty and vulnerability because of a lack of assets or outside income-producing opportunities, all of which is compounded by the absence of any social safety net. For instance, even in years with good rainfall there can be long-lasting hungry seasons rife with high food prices, malnutrition, and debilitating diseases like diarrhea and malaria. Any available employment opportunities, scarce at best, are for low-paid agricultural labour that can be taken up only by neglecting the household's family farm, resulting in low-yielding harvests and further under-production in future years.

These conditions are relevant to small businesses and informal enterprises as well, since these businesses are for the most part extremely vulnerable in disaster-prone areas. This is important to consider, because small and medium enterprises (SME) account for a third of all employment in low-income countries worldwide, and when added to smallholder farms and informal enterprises occupy the bulk of the labour force in many parts of Asia and Africa.

Compared with global businesses, informal sector producers and SMEs are far less resilient, particularly in low- and middle-income countries. Smaller businesses are more likely to be located in hazardous areas with evolving extensive risk, and are more likely to be wiped out entirely by a single disaster, whereas large companies, operating globally or regionally, diversify their holdings and can therefore recover from heavy losses. This is true even though they may suffer disruptions of their supply chains due to the incapacitation of some smaller companies that work with them. And while large, global enterprises may have much greater economic weight than SMEs, they do not employ many people proportionately. SMEs and the informal economy, on the other hand, represent much more employment. Yet they have an exceedingly limited capacity to recover from disasters compared with the large multi-nationals, and this has a negative impact on their survival rate. Small and informal businesses, therefore, are caught up in the disaster risk-poverty nexus along with poor households.

We have seen, in general, how the disaster risk-poverty nexus exacerbates both sides of the equation, with poverty increasing and/or creating disaster risk while disasters increase poverty. But even though most disasters are associated with extreme meteorological events, the social construction of disaster risk and the disaster risk-poverty nexus are not 'natural' phenomena. Both are the expression of historical social and economic relations that can be investigated from the point of view of their structure and their history. To put it simply, poverty is not a natural phenomenon, nor is its interaction with disaster risk beyond human intervention. The three published GARs did not, however, address a key question, namely, "Where does poverty come from?" That question has to be asked and at least provisionally answered if we are to find ways really to diminish disaster risk.

A great many points were brought out in the three reports supporting the view that disasters are endogenous to our socio-economic system, and if DRR efforts are to have any hope of genuine success, we must look at the way that system works. Since poverty is an abiding feature of our form of economic practice, and because its increase is a significant outgrowth of burgeoning inequality, we need to look at some of the salient features of this system and at the way they shape the parameters of disaster risk and the possibilities for its reduction.

Chapter 6
Global Capitalism and Disaster Risk

Since the disintegration of the Soviet Union, capitalism has had virtually unfettered access to the entire world. It is legitimate, therefore, to talk of global capitalism when we consider the economic practices that currently define our material life. While it would be impossible to give anything here but a misleadingly cryptic view of how capitalism operates, there are nonetheless a few general points that should be made that will help orient our analysis.

Capitalism is founded on the principle of economic growth and is structured so that growth is the engine of the system. It is also based on the private accumulation of value, expressed as money, which in this system is the ultimate purpose of growth. Evolving out of systems of exchange and simple markets, capitalism emerged as a qualitative shift through which the purpose of material production became the accumulation of money.

Exchange relationships date back thousands of years to the time when an economic surplus was first generated, i.e., when useful material goods began to be produced above and beyond what was needed for a community to survive and reproduce itself. This was primarily the result of the emergence of agriculture and the development of agricultural tools through trial and error, i.e., improvements in the forces of the production of material life.

When communities with diverse surpluses met, they exchanged these surpluses with each other on the basis of direct barter and according to what each community deemed useful to itself. There are many theories about how this barter operated, but the most coherent one is that there was some rough-and-ready sense of how long it took each community to make, grow, or gather the objects being exchanged. This calculation, then, became the foundation of each object's exchange value, the value it possessed solely as an object of exchange.

This exchange value must be distinguished from an object's use value, i.e., the value it has in being perceived as useful by someone. Looked at from another angle,

M. Gordy, *Disaster Risk Reduction and the Global System*,
SpringerBriefs in Climate Studies, DOI 10.1007/978-3-319-41667-0_6

exchange value is a function of a social relationship between two or more people, while use value can refer simply to a single individual. And while in principle every object with exchange value has to have, or be perceived to have, some use value as well, an object with use value may have no exchange value at all. We distinguish objects possessing exchange value from objects per se by calling them *commodities*.

As social relationships of exchange became more widespread and complex, and as more objects became commodities, markets emerged where multiple exchanges took place. Eventually, one commodity in each market emerged as the medium of exchange, in terms of which the exchange value of all other commodities was calculated. Whether this single commodity was gold, silver, beads, furs, shells, or what-have-you, exchanges were facilitated by having it serve this function. We call such a medium of exchange *money*.

After some time, individuals appeared in the market who had somehow accumulated enough money so that they had more than they needed to provide themselves with the commodities sufficient to satisfy their material needs and desires. Yet money is social power, i.e., power over the labour of others inasmuch as it can purchase commodities that are the output of somebody's work, and this power is in itself something that people can and often do desire, no matter how much they may already have. The individuals in question thus proceeded to *invest* their money with the intent of accumulating more of it in the process. Their interest was not in production as such nor in the products produced, except insofar as these products could be turned into more money than they had invested, i.e., insofar as they made a *profit*. The primary interest of these investors, and often the only one, was (and is!) monetary profit.

Heretofore, market participants had come to exchange commodities for money so they could buy other commodities, ultimately because of the use value these commodities had rather than their exchange value. Now, however, some people came to the market with money to invest in producing commodities, with the goal not of obtaining products but of accumulating money.

The aforementioned investors, both then and now, purchase infrastructure and materials. They also hire workers, whose wages are a cost of production, to work with this infrastructure and these materials to produce commodities which they then hope to sell for more than the sum total of what they have invested. This monetary increase, this profit, is calculated as a percentage of the entire investment. For instance, a product that costs $100 to make, all costs included, and is then sold for $110, will have generated an absolute gain of $10, or a ten percent profit. If all the proceeds of this process are then reinvested, i.e., if the entire $110 is put back into a new production cycle, then the output must generate an increase of $11 in order to maintain the same profit margin of 10 %. It is in the interest of maintaining or increasing this rate of profit that capitalists try to keep the costs of production as low as possible.

Since markets are built on the principle of equal exchange, i.e., all commodities in any specific exchange are considered to be of equal value, profit generation seems to be mysterious because in principle all the elements of the production process, including the power of labour, are purchased at their fair market value. Since the selling price of the product is supposed to be the sum of all these inputs, this means that in principle the product cannot have more value than was put in at the outset except for what was added by the workers who produced it. The capitalist, of course, could pay himself a wage for organizing the business, although this falls outside his role as a pure investor and merely adds to the cost of labour.

While the solution to this mystery remains controversial, it should be kept in mind that no capitalist, acting strictly in his capacity as capitalist, ever hires anyone if he doesn't believe that this person will create more value by his work than he is paid in wages. (When some business owner hires his ne'er-do-well nephew as a favour to his sister, he is not acting as a capitalist.)

Money that is allocated to this process, or more generally, money that is allocated to making more money in any way, is called *capital*. Capital thus implies *increase*. Under this definition, capital and investment are two sides of the same coin, and investment should be distinguished from allocating resources, monetary or otherwise, for other purposes. For the moment, however, we will continue to call all these other allocations investments out of respect for common parlance, although speaking of 'public investment' is a bit misleading, as is using the term 'natural capital'.

It is important to note in passing that in this production process the social relationship between the capitalist/investor and labour is unbalanced. The capitalist, possessing money to invest, comes out of each production cycle with more money than when he went into it; this is his profit. The labourer, however, does not accumulate because he does not share in the profit. He gets paid for his time, which is all he has to sell, and at a rate that only allows him to reproduce his existence as hired help. Keeping in mind that money is social power, the labourer's relative social standing is increasingly diminished because of these unbalanced social relations of production. This is a systemic engine of increasing inequality between the relatively few capitalists and the mass of the working population, and while it is not the only such engine, it is certainly the most powerful because it is in fact systemic. Since increasing economic inequality is clearly a barrier to poverty alleviation, this systemic imbalance feeds what has been identified as a very significant underlying disaster risk factor.

It is clear from what has just been said that maintaining a stable rate of profit requires an absolute increase in production. That in itself will not guarantee capitalist economic growth, however. The products must be sold in the market and turned back into money, and that entails something called effective demand. There must be people in the market who want the product and, most importantly, have the money needed to buy it at the market price, i.e., their demand must be *effective*.

A problem eventually arises when markets are saturated, i.e., when the need for expanded production outstrips the ability of the market to absorb, leaving the capitalist with unsold products and generating a fall in his rate of profit. Since the need for capital to expand is calculated by a multiplier, a percentage rate of profit, while the ability of a market to expand, or to be expanded, is arithmetic, i.e., additions to it are made one-by-one and not as a multiple, a contradiction arises.

Existing markets must be deepened or new markets must be found, usually in competition with other producers. And these markets must be populated with buyers who have money in their pockets.

History is replete with examples of how capitalists, individually or in groups, work to expand their markets, but here is not the place to recount all this. It suffices to say that this expansion has to take place or capitalism would crumble under the weight of what is known as a crisis of effective demand, an endemic characteristic of the system that gave birth to the slogan, "Grow or die."

Although the foregoing account may seem simplistic to the point of simple-mindedness, nonetheless it raises several points that will eventually enlighten us about disaster generation and the possibilities for significantly reducing disaster risk. Since the vast majority of resource allocation throughout the world consists of private investment, and since the purpose of that investment is to make more money for the investors (the sacred obligation to 'enhance shareholder value'), how is it possible to interest elements of the private sector in supporting DRR when doing so might limit their profit making? How can you convince capitalists, acting as capitalists, to desist from constructing a nuclear power plant on a fault line when the anticipated profits from doing so are enticing, especially during times of market saturation? Let's look at this a bit more closely.

Human beings are capable of any kind of behaviour, from the most noble to the most reprehensible, and history has examples of any and every behaviour one can imagine. However, human beings are social animals; every human characteristic is a product of social interaction, and human physical survival depends on being born into a social structure of some sort, even if it is only that of a single mother taking care of her child.

Every social structure encourages certain behaviours and represses others, whether explicitly through coercion or softly through formal and informal education. In a social environment where the most important social practice, economic practice, is built on the insatiable drive for private capital accumulation, it seems reasonable to expect that other values will recede in importance as time passes. A relevant example here is the phenomenon that in countries where public infrastructure and services have been privatized, incorporating DRR statistics and concerns in the investment strategies for those services has become more than challenging.

Acting according to the dominant economic paradigm comes to seem 'natural', a part of 'human nature', which is then paraded as a justification for what are, after all, social relations of a particular type and historical origin. This, then, is offered as support for the view that the only viable responses to burgeoning disaster risk must be ones that leave the current social relations of production intact, namely, so-called "market-based solutions". Carbon trading, geo-engineering, new business opportunities for the insurance sector, etc., are examples of trying to respond to the problems caused by 'business as usual' with…business as usual. Anything else tends to be seen by the investment community as 'unrealistic'.

That investment community is becoming increasingly trans-national, a phenomenon driven in part by the search for new markets and for the investment opportunities offered by opening up previously non-capitalist countries to invest-ment from everywhere via trade agreements. This evolution was facilitated by scrapping virtually all regulation of capital movements in the late 1980s and 1990s, along with regime change in countries that were previously nominally socialist. Helpful as well were innovations in telecommunications technology that made it possible to move money around the world at the speed of light, enabling the financial sector to make money out of moving money. In fact, profits increased so substantially in the financial sector that finance became the dominant fraction of capital, underwriting the penetration of private investment into all corners of the globe.

Multinational companies are now comprised of investors from everywhere and have set up production facilities wherever labour costs and taxes are lowest and environmental constraints most flexible. Ties between global corporations and any particular country have been weakened to the point where it is nearly impossible to say whether an enterprise is British or Chinese or Japanese or American, regardless of its legal domiciliation. This has made possible a certain corporate indifference to location, so profit making can now proceed with scant concern about where or why or how it takes place, except inasmuch as the image of corporate social responsi-bility must be maintained for public relations purposes.

Something similar has been happening in the financial sector as a whole. Global deregulation of that sector catalyzed a staggering increase in the volume of capital associated with speculation and short-term investment, with the total value of equity market capitalization and outstanding bonds and loans reaching US$212 trillion by the end of 2010. At the same time, financial capital has been concentrated in a limited number of large institutions and a largely unregulated 'shadow banking' system. This shadow banking system is a complex value chain of intermediaries, including investment banks, hedge funds, and equity funds, that enables assets to be moved around the world using a large number of financial instruments that facilitate investment in physical assets as well as in production and services.

Through this system, the financial sector has gained the ability to respond and adapt quickly to short-term investment opportunities, although this rapidity has made it much more difficult to account for longer-term risks and liabilities in investment decisions or for systemic risks such as financial bubbles and the like. So when risks are considered in short-term and speculative investment decisions, disaster risk rarely figures in the deliberations. The increasing sophistication, complexity, and opaqueness of financial instruments means that securities and bonds for businesses with a high level of disaster risk are bought and sold without ever considering how these risks may affect asset values, let alone how they may impact on the lives and well-being of local populations.

Many of these investments are hidden behind complex and aggregated corporate loans, financed by large funds or banks whose individual investors rarely know what specific activities are being financed. Beneficiaries in high income countries with well-established pension funds may unknowingly be benefitting from disaster risk transfer, for example by investing in agricultural practices that increase drought risk. Likewise, investors tend to underestimate or ignore systemic risk, in part because they are so far removed from the site of the actual investment. As the cliché goes about love, so is it true about investments made at far remove: "Absence makes the heart grow fonder."

The massive increases in productivity and monetary growth achieved by this deregulated system over the past forty years have been matched by an equally massive increase in shared risk. This is risk that is passed from the primary beneficiaries of economic expansion to the population as a whole, most of whom do not share in the economic benefits. Because investors have placed a premium on short-term profits rather than on sustainability and resilience, disaster risk has increased substantially, especially in areas where capital flows are volatile, i.e., characterized by so-called 'hot money'.

Businesses have exploited the comparative advantages of different locations by decentralizing and outsourcing production while accelerating turnover time by fine-tuning and diversifying their supply chains. This increases their exposure to earthquakes, storms, tsunamis, floods, and droughts, because many of the locations they choose for their investments are hazard-prone. In addition, quite a number of these investments are rapidly turned over under the pressure of capital volatility and other exigencies of the global financial system, which decreases investor concern about longer-term consequences. Faced with currency fluctuations, global financial crises, or pressure from large financial institutions for increased payback, businesses often pay insufficient attention to disaster risk or to sustainability, environmental capacities, and resilience when investment decisions are made.

The risks associated with this inattention are passed on to the next owners and to the surrounding communities. Governments, who have often underwritten significant parts of the necessary social infrastructure in order to win the competition for these investments, are frequently left with liabilities to amortize when disaster strikes. Again, this is a case of privatized benefits and socialized costs.

Businesses do, however, take care of their own disaster risk interests if they are large enough and powerful enough to do so, mainly under the rubric of business continuity planning. This approach enables them to identify and anticipate threats to their operations and supply chains by designing contingency plans that may include rebuilding elsewhere, which will allow their operations to be resumed quickly and with minimal disruption. While that may be a useful tactic for them, it needs to be part of a strategy that takes into account the way their investment decisions are modifying the level of the disaster risk they face. Ideally, they would also try to anticipate the impact of their investments on the disaster risks, often the same ones, facing the local population, and on the sustainability not only of their own commercial activities but on the viability of the local economies where they operate. This would be a way of showing that the concept of corporate social responsibility has some practical meaning above and beyond image enhancement, but the systemic law of private accumulation and the relentless struggle to maximize profit work against this.

The point was made in the GARs, especially in GAR 2013, that investing in disaster risk management has important benefits for businesses. These include less uncertainty, along with greater confidence among investors that risks are being anticipated. There are also cost reductions associated with avoiding business disruption. Escaping the expense and disorganization of rehabilitating or relocating damaged plant and other facilities is another advantage, along with the relative cost savings associated with taking anticipatory or preventive action.

On the other hand, throughout the GARs there are many indications that the truly risk-generating behaviours of the private sector, such as building production facilities and housing in hazardous locations, investing in short-sighted and self-destructive agricultural practices, expanding the use of fossil fuels, limiting expenditure on resilience, promoting wasteful consumption, and calculating the safety of the surrounding population against the bottom line, among many others, will not recede significantly unless the paradigm of unbridled private accumulation is overcome. Since this paradigm is written into the system as a whole, either the system must be transformed radically or else it must be tamed by strong regulation based on the value of disaster reduction for everyone, i.e., placing the well-being of humanity ahead of private profit. To come to grips seriously with the problem of accelerating disaster risk, no "market-based solutions" will ever be sufficient.

Maintaining the system as it is, however self-destructive that may be, depends on government reproducing the system's conditions of existence. On the other hand, effecting the necessary changes to that system also depends on governments, or at least on political interventions, in this case reflecting pressure from the mass of the populations that governments are meant to represent. This is the governance choice *sine qua non,* and it is a topic to which we now turn.

Chapter 7
Political Will and Governance

Private enterprise is subject to the inexorable pressure of having to make a profit above all else. Businesses that fail in this do not survive, a structural constraint that makes it virtually impossible for the private sector to take a neutral view towards being regulated, especially if a regulation limits even its short-term profit-making possibilities. The question of whether the regulation in question is in everyone's interest, including its own, does not enter into its deliberations.

Because it is constrained to see everything through the prism of private profit and loss, the business sector cannot see DRR except in financial terms first and foremost, even calculating mortality in these terms. Although many business people may not subscribe individually to this view, the resulting de-personalization of human suffering becomes an institutionalized, if to some degree unconscious, feature of business calculation.

Governments, on the other hand, have an option. Unconstrained by the constant need to make a profit, they are not obliged to subscribe to the business accounting perspective. This leaves open the possibility that governments can be induced to act in the interests of the population as a whole with regard to DRR and to try and alleviate some of the more egregious elements of business practice noted in the previous section. They can in principle, if so inspired, put the well-being of the entire population ahead of the financial profits of a minority.

Yet even though governments are not formally subject to the laws of private capital accumulation, in practice they act too frequently as if they are. In most countries there is a symbiotic relationship between governments and the groups at the pinnacle of economic power. That relationship needs to be loosened if meaningful regulation of business investment is to take place, and the most promising means of achieving this lies with collective action at the community level.

Surveys taken across the DRR sector in recent years have asked for suggestions about what to emphasize in an extended, post-2015 version of the Hyogo

© The Author(s) 2016
M. Gordy, *Disaster Risk Reduction and the Global System*,
SpringerBriefs in Climate Studies, DOI 10.1007/978-3-319-41667-0_7

Framework for Action (HFA). Overwhelmingly, respondents said that DRR should become more solidly anchored in the community, at the grassroots level, and that community organizations should be more strongly supported in their efforts to promote resilience. This is because communities are always in the front lines when disasters strike, and are most strongly and immediately affected.

In principle, decentralizing disaster risk management (DRM) functions to empower communities facilitates citizen participation, inspires more engagement among decision makers, makes better use of local knowledge, uses available resources more effectively, and strengthens accountability. While this principle may not always be fully realized, it provides a regulative ideal towards which it pays to strive, since any progress in this direction enhances social cohesion and can lead to greater investment in DRR.

On a daily basis, local communities tend to be less beholden to the power of trans-national capital and its dominant global economic paradigm than central governments, and while there is no doubt that self-interested politics play an important role at the local level, those politics are often more sensitive to local conditions and to the views of the community as a whole. It is more difficult to ignore the needs of one's neighbours than it is to view their concerns with relative indifference from the national capital or from a boardroom on the other side of the planet.

To make a practical difference, this shift in focus requires a significant transfer of DRM resources to the local level, otherwise genuine empowerment cannot occur. Moving responsibility to local authorities and organizations without sufficient funding is merely abdication. Political decisions have to be made at the national level to devolve real power over DRM to communities.

It is clear that political decisions about massive transfers of public resources do occur concerning issues other than devolution of DRM; witness the political response to financial crises. During such crises, central governments act quickly to provide money to shore up banking systems and protect wealth. In a more rational world, human mortality and the destruction of the livelihoods and futures of the broad mass of the population should inspire similar actions. In fact, governments often say that a lack of financial resources constrains such investment. Nonetheless, how available resources are allocated reflects political priorities, and in most cases governments can allocate a greater proportion of public money to DRM than they currently do.

The reason that they don't is a matter of political weight and pressure, which is why an effective politics for DRR and DRM has to begin at the community level. Because local politics are somewhat more sensitive to the wishes and needs of the whole population, non-governmental movements need to be encouraged to pressure local authorities to take positions that in some situations go against the policies and plans of the central governments and trans-national capital, for example with regard to hydraulic fracturing gas extraction (fracking).

One reason this community mobilization is required is that so many central governments are in thrall to powerful private interests in sectors such as urban development, construction, agribusiness, and tourism. These interests may offer strong disincentives for investing in DRM. For instance, agribusiness may want to privatize water resources, and governments may see this as a way to increase agricultural productivity and generate foreign exchange. The result, however, is to transfer agricultural drought risk to subsistence farmers. This is because privatizing water resources may lead to their overuse by wealthy agribusiness enterprises, exacerbating scarcity and invariably raising their market price. That in turn squeezes small holdings that haven't the financial resources to buy the water they need, converting agricultural drought into a disaster for them. It also makes it much more difficult to regulate water use, as was mentioned above with regard to privatizing public services generally. Effective DRM would require keeping water resources, as well as many other public services, entirely within the public sphere, benefitting rich and poor alike.

The people most in need of government help in this regard, poor farmers and marginalized portions of the community, have a very difficult time making their concerns heard in the upper reaches of political power. On their own they are rarely in a position to do so. The pressure required to break this stranglehold must therefore begin with collective action among various communities, in concert with expert institutions and civil society organizations that include some NGOs. Networked communities and their allies can create leverage, first of all to bring the concerns of the marginalized strongly into governmental focus, and even more importantly to organize local initiatives and generate government support for them. Sometimes the space for these initiatives must be carved out by organizing community-network resistance to economic projects that add to disaster risk, but the overall purpose is not negative. The point is to substitute risk-neutral or even risk-mitigating development for risk-generating economic activities.

This will entail a shift in perspective at the community level concerning regulation of private investment. The belief system promoted by global capital has increasingly penetrated the consciousness of people all around the world, at least until fairly recently, and dominates the thinking in rich countries as well as in the rich sectors of countries that are not rich. It holds that virtually all regulation of business is bad not only for business profits but, through them, for society as a whole. GDP growth, which is basically a measure of private capital accumulation, is taken to be the gauge of a country's economic health. Anything that limits GDP growth, especially government regulation, must be avoided at all costs so that 'the miracle of the marketplace' can do its work unfettered.

That does not mean that private enterprise is against all regulation. Only regulation imposed from outside the private sector is to be eschewed. Voluntary regulatory frameworks are acceptable because they have been devised by the companies that are willingly submitting to their jurisdiction. Since the interests of private capital accumulation are stipulated to be harmonious with the good of the entire community, if not identical with them, business self-regulation creates the best of all possible worlds. Any problems are thus taken to be exogenous, including disasters.

This has been the prevailing ideology since the 1970s, but it is quickly losing its lustre. In any case, outside the dominant economic and political circles across the globe it was never wholeheartedly accepted by most people except insofar as they harboured dreams of becoming rich and powerful themselves. In recent years, of course, the 2008 financial meltdown, which has become an economic and social meltdown as well, has brought the concept of government regulation of business out of the shadows. The more general principle of government intervention in the marketplace was revived through the myriad of ongoing bailouts of moribund financial institutions, along with government assurance of their profits and bonuses, and so many people are now asking why regulation cannot, in fact, be imposed. Nonetheless, the ideology of the 'free' market still forms a serious obstacle to effective government action, as do the after-effects of policies installed during its hegemonic period.

For instance, among many low- and middle-income countries, so-called structural adjustment programmes, imposed as a condition for IMF loans, were used as a vehicle to compel governments to remove all regulatory barriers to investment and growth; to reduce government spending on social services and health; and to ensure debt servicing. In order to compete for investment, governments have done these same things on their own initiative and have in addition sold off public industries and services to private investors. Governments have thereby taken on the role of business promoters and facilitators of private investment.

Financial, labour, and property markets were deregulated, tariffs on trade were reduced or eliminated, and incentives were arranged to attract business investment. In some countries, national institutions for development planning were either weakened or closed down. In such circumstances it was difficult to see government as an 'honest broker' between competing interests, because it acted explicitly as the handmaiden of private accumulation no matter what the cost to the welfare and safety of the general population, especially of the most vulnerable and politically powerless.

Despite the growing acceptance of the need for governments to regulate private investment, countries have not been very successful in achieving economic risk-sensitivity, even though national governments and local administrations are best placed to devise and implement investment frameworks. That is why political and social pressure from the grassroots is needed in order to focus government attention on risk-generating development and to provide political incentives for government to play its necessary role in addressing the underlying drivers of disaster risk. Without this kind of pressure, the field is left open for private enterprise to make its case unanswered and for national policies, institutional frameworks, and legislation on DRM to remain largely peripheral in the struggle to weaken those drivers.

In short, legislative frameworks and the like are little more than weak palliatives if they are not accompanied by genuine implementation and rigorous enforcement. However, governments have not provided adequate disincentives to the kind of business investment that generates disaster risk. In this context, organized community pressure offers the best possibility of holding governments accountable for what they have committed themselves to do, and to counterbalance pressures from powerful economic interests aimed at making sure that laws mandating risk-neutral investment remain toothless.

This means that the political will of the community must become the political will of the government. Since the incentives for disaster risk reduction are strongest at the community level, it is possible for popular pressure to arise even though many people may initially feel, on the basis of their lived experience, that nothing can be done. The first thing necessary for this to come about is to make extensive disaster risk visible and recognize its importance, for it is extensive risk that affects communities, especially poor ones, most directly in their daily lives.

Chapter 8
The Importance of Extensive Risk

Extensive disasters are mostly associated with highly localized, recurring hazard events such as flooding, landslides, fires, and storms that are considered to be small in scale. How they are reported depends on the perspective of the observer, which means in practice that extensive disasters are largely invisible at the global level and not often reported nationally. Nonetheless, these disasters are much more frequent than intensive ones and have impacts that are in some ways even more dangerous over time. In addition, most of them are connected to weather-related hazards.

Statistical sampling in the GARs indicates that the annual occurrence of weather-related disasters has doubled over the past 30 years, although economic losses are growing faster than mortality, reflecting the increasing exposure of assets and the geographical expansion of weather-related risks. This exposure is due in part to economic development, with more valuable assets being constructed in hazard-prone areas, for example tourism infrastructure in SIDS or expanded informal urban settlements in flood plains or on landslide-ridden hillsides.

Although extensive disaster risks have only recently begun to be taken seriously, there is nonetheless anecdotal evidence to support the suspicion that mortality associated with extensive disasters is being under-counted. Destruction of economic infrastructure has long-term effects on poverty, which underlies all sorts of health problems that ultimately increase death rates and morbidity. The true costs of extensive risks for human life, while difficult to ascertain, nonetheless need to be integrated with the mortality totals associated with extensive as well as intensive disasters.

For example, recurring extensive disasters like extreme drought or flooding can destroy food security to the extent that extreme hunger increases in a community. Deaths from starvation are not considered to be the direct result of these disasters, but they are real nonetheless and would not have occurred if the disasters had not taken place. The same accounting problem faces intensive disasters. At Fukushima,

© The Author(s) 2016
M. Gordy, *Disaster Risk Reduction and the Global System*,
SpringerBriefs in Climate Studies, DOI 10.1007/978-3-319-41667-0_8

for example, deaths from radiation exposure and its associated cancers will not necessarily all appear in the short run, nor will mortality from food sources poisoned by the nuclear meltdown. But appear they will at some point, with startling increases in disease-related deaths. Certainly a new accounting method needs to be devised that will synthesize economic loss with mortality and will aggregate immediate effects with ones occurring in the medium- and longer-terms.

Overall, the costs associated with extensive disasters are significant in relation to the communities where they occur. The direct economic costs of assets in poor communities, calculated monetarily, may seem low compared to economic losses in richer ones, but the difficulties in replacing these assets are much more pronounced in the former. Coupled with the accumulated follow-on mortality mentioned earlier, the loss of livestock and other productive assets can devastate rural livelihoods, crippling recovery and feeding back into increased poverty and vulnerability in the face of more frequent localized extreme weather events. In poor urban areas, the loss of a house can also have truly catastrophic consequences, especially when the house is the locus of the productive or commercial activities on which the household depends. Because, as we have seen, poor households generally have no buffer against the economic consequences of disasters of any kind, intensive or extensive, their losses far outweigh what can be expressed monetarily.

It should also be noted that while extensive disasters do not generate the level of mortality associated with intensive ones, they account for a large proportion of losses to public assets such as health and educational facilities and infrastructure, as well as for the loss of livelihoods mentioned earlier. These losses affect low-income groups most sharply, and because of the greater vulnerability of these groups, they can have a negative effect on mortality over the longer term. These indirect disaster losses are not often accounted for in risk assessments.

As new development decisions and investments interact with disaster risk, they have impacts which may not be immediately apparent. It may be years, or even decades, before these impacts manifest in loss of life, ruined livelihoods, or damaged infrastructure. As long as these losses go unmanaged, they may have further effects such as increased poverty, declining human development, and reduced economic growth. Since the vast majority of these impacts are extensive in character and accumulate over time, occurring throughout a country or region, they manifest as a large and rising number of localized disasters, mainly associated with storms, floods, fires, and landslides, all of which are connected to climate variability.

Since extensive disasters have a tendency to recur in areas subject to severe levels of similar kinds of risks, over time extensive risk can become intensive. Accumulated losses can reach a critical mass and plunge the community into a downward spiral. Repeated disasters wear down the morale and asset reserves of local populations to the point where they have a very hard time putting together any kind of recovery effort.

Seen from this angle, extensive risk represents the initial stage of disaster risk accumulation and should be addressed as a matter of primary importance. Doing that would normally involve relatively small investments, for example improving storm drainage in informal settlements. It is therefore cost effective to deal with extensive risk now rather than having to respond to intensive disasters later on.

Most important, extensive risk must not be allowed to accumulate and increase poverty. Poverty, as we have noted, is a crucial underlying factor for disaster risk and enhances vulnerability. Allowing extensive risk to accumulate thus strengthens an important risk driver and this in itself increases the danger. The vicious circle has to be broken.

Extensive risks can also contribute to the impacts of intensive disasters. For example, local flooding can greatly increase the mortality and economic loss associated with an earthquake, which has been the case in Dhaka, Bangladesh, an urban area of over 15 million people. Many areas surrounding central Dhaka are flood prone during the rainy season, and until recently were occupied by natural water bodies and drains, vital for flood regulation. While in principle limited by land use planning instruments, private and public sector projects are nonetheless rapidly urbanizing these areas, and this removes these natural flood regulators. Destroying retention ponds and drains increases flood risk just as building in drained wetlands increases earthquake risk. During an earthquake, sand and silt can liquefy to the point where the ground can no longer support the weight of buildings, which then often collapse or suffer heavy damage. The processes that configure risk are therefore very complex, posing serious challenges to effective disaster risk governance.

Because extensive risks have not been adequately included in disaster risk assessments, these challenges have not been fully addressed. More significantly, since extensive risks turn into disasters more readily in poor communities, those communities have been marginalized in DRR planning. While this is slowly beginning to change, communities themselves need to make their circumstances known more clearly at all governmental levels. This can happen only if communities become aware that extensive disasters are not simply an unavoidable part of their lives and that many social conditions contribute to their being more vulnerable and exposed to risk. Then they can organize political pressure on governments to facilitate their self-help by allocating resources to local initiatives that are run by the communities themselves.

Local awareness of how extensive risk can be diminished, accompanied by community political organizing based on that awareness, will not by themselves ameliorate the complex problems associated with DRR. They can, however, contribute by at least bringing extensive disasters into the light of national and international public scrutiny. A helpful response to this from central governments will be required in order for community efforts to reach their potential, but making extensive risk and its impacts more visible beyond the populations directly affected is in itself a powerful political lever that must be used.

The visibility of extensive risk is one part of a larger issue, however, and that issue is the overall invisibility of some important aspects of disaster risk. We shall now examine this invisibility in light of what has been said so far about the dominant economic development paradigm and the conceptual framework associated with it.

Chapter 9
The Invisibility of Disaster Risks

Earlier we mentioned that economic losses from disasters might seem low in poor communities and that this was misleading. That is because the calculation of loss is often confined to the financial costs of replacing what was damaged or destroyed. The low financial value of assets in poor communities, however, is a reflection of the impoverished quality of housing, infrastructure, and services characteristic of these communities and understates the negative impacts of their loss on poor households. To a large extent, an adequate calculation cannot be restricted to a purely financial evaluation. Asset losses among the poor are therefore difficult to measure according to what, in the last instance, is a business paradigm.

Take once again the example of the loss of urban housing and what this means to a household's future economic opportunities. Particularly in informal settlements, the house is a production site and is often also a centre of commercial activity. Its loss is a severe blow to any household that depends on what it produces and sells to survive. Even if it is sometimes possible for a household to reconstitute its economic life elsewhere, the disruption of its work and the time lost in making that transition can be a blow to its livelihood from which it may never fully recover. And given their low status and lack of secure tenure, households in informal settlements are generally excluded from public investments in vital risk-reducing infrastructure and services.

Likewise, monetary undervaluation of ecosystem services acts as an important obstacle to adopting eco-system DRM. Eco-systems are often taken for granted as relatively 'free' natural capital because these systems are not 'produced' by human beings. They are treated as something 'found' in nature, much as water sources used to be (and oxygen in the air still is).

Because eco-systems intersect with various economic activities, it is unusual for their value to be calculated in other than monetary terms. Yet it is difficult to put a monetary notation on mangrove forests, for instance, that adequately reflects their

© The Author(s) 2016
M. Gordy, *Disaster Risk Reduction and the Global System*,
SpringerBriefs in Climate Studies, DOI 10.1007/978-3-319-41667-0_9

multi-faceted value to DRR and therefore to protecting those economic activities. Risks to them are therefore under-appreciated and remain more-or-less invisible until disaster strikes. When that happens, losses are calculated for the assets the eco-system was meant to protect rather than for the eco-systems themselves. As a consequence, few countries are taking advantage of tools such as 'payments for ecosystem services'.

There are other examples. For instance, the lack of systematic recording of drought losses and impacts, particularly insofar as they affect poor and vulnerable rural households, contributes to their reduced economic and political visibility. This is reflected, in turn, in the weak motivation in many governments to do anything about the underlying factors that drive drought risk, a weakness demonstrated by policies that promote tourism and urban development in water-scarce areas.

Sometimes undervaluation is deliberate. When there are structures in place such as conditional cash transfers to offset a driver of disaster risk like structural poverty, there is often an incentive to set the poverty line too low in order to reduce the cost of poverty reduction programmes. This reduction, of course, limits the scope and effectiveness of the programmes and thus limits their impact on an underlying disaster risk driver. Since poverty and disaster risk form an unholy, self-reinforcing alliance, the overall costs to society (including mortality) simply increase. In addition, by not counting sections of the poor as truly poverty-stricken, this practice remains short-sighted in another sense because it renders them administratively invisible.

Setting the poverty line too low is often said to be a characteristic of the governments of low-to-medium income countries, especially those that have had structural adjustment programmes imposed on them as a condition for essential loans. However, such under-evaluation is becoming common in rich countries as well, as a side effect of the domination of neo-liberal ideology in the prevailing socio-economic paradigm. There the notion of the 'undeserving poor' has taken hold in political parlance in the interest of cutting back state expenditures on social needs, and the calculation of what constitutes poverty reflects that notion. As a pundit once put it, "The rich don't want even the crumbs to fall off their plates."

In the previous section we noted that extensive risk remains largely invisible at the international level and implied that at the national level the losses associated with extensive disasters are often under-counted, both in terms of assets and mortality. It may be that there is a multi-dimensional set of interactions between this lack of visibility, the social construction of risk, extensive disaster risk, and the poverty-disaster risk nexus.

It is clear that poor populations are much more vulnerable to disasters than richer ones, even when the hazards are the same. Unable to provide themselves with resilient buildings and infrastructure, and often incapable of accessing emergency services, poor communities are subject to the full force of whatever extreme events occur, which turns these events into full-fledged disasters for them. Since the poverty of these communities and the relative indifference of national authorities to their vulnerabilities are the result of a history of a certain type of social relations, there is reason to recognize this poverty-disaster connection as an example of the social construction of risk.

More specifically, poor communities are much more vulnerable and often more exposed to extensive risk than communities that are materially better off. Poor populations are more likely to be located in hazardous and less desirable areas, and may also find they are living next to land where dangerous and/or polluting economic activities are taking place. Informal communities built on flood plains or on denuded hillsides are more likely to suffer from frequent floods or landslides, while unregulated manufacturing close by may generate all sorts of hazards such as chemical fires and other sources of toxicity.

Having no choice, poor communities accept these extensive disasters as an unavoidable part of daily life. This resignation is a sign of political marginalization, but it is a partial cause of that marginalization as well. And political marginalization renders poor people and their concerns less visible, which in turn has an impact on the visibility of extensive disaster risk, the kind of risk that affects the poor most strongly.

Another factor in risk invisibility has to do with the way disaster losses are generally calculated. It is clearly easier to account for asset losses that are insured than it is for those that are not. This is due, in part, to the fact that insurers and re-insurers maintain extensive data on insured losses, and they do this because it is essential to their businesses. These data can then be accessed by governments through arrangements with insurance companies and can inform DRM.

However, most insured losses pertain to intensive disasters, and this can be explained in part by the fact that the largest proportion of extensive disaster risk falls on poor populations, as indicated earlier. These populations do not have the wherewithal to insure against intensive disasters, let alone extensive ones, so their losses are easily ignored outside of the communities affected, although they may be noticed to some degree at the national level. The implication of all this is that the headline-grabbing figures recorded in global datasets over the last decade may be quite conservative. Once the losses associated with nationally reported smaller disasters are included, incomplete though their accounting may be, those figures are likely to be at least 50 % higher.

Even more significantly, the indirect losses and the wider effects of both intensive and extensive disasters are rarely considered, especially when they pertain mainly to low-income households and communities. These comprise negative welfare outcomes that include declines in school attendance, nutrition, health, productivity, and increases in inequality and unemployment, outcomes that can be transmitted across generations. Social costs like these, particularly when they stem from extensive risks, are not taken into account by either governments or business and are therefore absorbed directly by low-income households and communities whose development potential is thus undermined and whose resilience is eroded.

Another source of risk blindness is discounting systemic risks, particularly disaster risks, in investment decision-making. Investment decisions that may seem rational from the point of view of the individual investor may add to the fragility of the socio-economic system and in the end turn out to be irrational. Such was the case in the housing bubble that burst in 2008 and the same may be true for disaster risk-generating individual investments over the long run, both in terms of straightforward asset loss and of the aggregated social and environmental impacts on human beings.

Because their fiduciary responsibility is primarily to identify investment opportunities, asset managers rarely consider disaster risk when making their investments. The increasing distance between these managers and their beneficiaries, resulting from the financialization of the world economy, increases their disconnect from the way the invested money is ultimately used. Add to this the opacity and sophistication of contemporary financial instruments and you have a recipe for obliviousness.

Even when disaster risks are recognized, the decision to invest in a particular asset may nonetheless be made because the anticipated profits outweigh any countervailing risks. This is so because quite frequently a significant portion of these risks can be transferred to the public sector and to other countries and regions. From the business perspective, then, these risks can be considered as externalities and are thus virtually invisible in this context. Yet as stated in GAR 2013, private profit-oriented investment in recent years may have resulted in 2011 being the year with the largest disaster-related economic and insurance losses ever (GAR 2013, p. 195).

The global financial crisis that began in 2007 and continues today was caused in part by an over-accumulation of financial risk from huge flows of capital into speculative, debt-financed urban development. This debt, and the risks it internalized, was then sold on and shared through opaque investment vehicles that had not been (or perhaps could not have been) assessed or valuated. The financial risk was thus rendered invisible.

The accumulation of disaster risk in recent decades is analogous. In many hazard-exposed countries, governments, institutional investors, businesses, and households are currently sitting on a mountain of hidden debt, namely, contingent liabilities represented by unrealized disaster risk. This disaster-prone capital stock, whether privately or publicly owned, represents another category of toxic assets that do not appear on any balance sheets but which may, in fact, be lethally as well as financially dangerous. Since disaster-prone capital is disproportionately tied to urban development, which rests in many cases on with speculative investment, we must now turn our attention to urbanization and land use issues that are relevant to reducing the underlying drivers of disaster risk.

Chapter 10
Urbanization and Land Use

In the first section of this document we saw that unbridled urbanization, especially in poor regions, is an important underlying disaster risk driver. We also had a brief glimpse of some of its effects. But what drives this kind of urbanization? One causal element has to do with free trade agreements between richer and poorer countries, an aspect of the trans-nationalization of capital over the past 40 years, and the increasing imbalance of political power relations between rich and poor in all countries and on a global scale. Unbridled urbanization and its accompanying misuse of land cannot, therefore, be treated in isolation from a variety of other factors.

The phenomena of unregulated and uncontrolled urbanization, rural immisera-tion, eco-system degradation, real estate speculation, globalization, and the disaster risk-enhancing practices of agribusiness are often noted, but the overall context of these phenomena is missing and the history of their development is nowhere to be found in the three published GARs.

That overall context may be summarized very briefly as follows: the imposition of free trade regimes by the North on the countries of the South, often through IMF structural adjustment programmes or occasionally by 'regime change', accelerated urbanization significantly by opening the markets of poor countries to an influx of subsidized agricultural products from the North that undercut small farmers and pushed them to destitution. Rural populations were thus forced off the land and migrated to cities in hopes of finding work, but mostly what they found was that the infrastructure, economic opportunities, and governance required to welcome them were non-existent. They were forced into informal lodging with no building stan-dards or safety provisions, while the land they vacated was taken over by inter-national agribusiness, whose lack of concern for issues of sustainability and environmental degradation is legendary. And while this dynamic pre-existed these trade agreements and structural adjustment programmes to some degree, its

© The Author(s) 2016

M. Gordy, *Disaster Risk Reduction and the Global System*,
SpringerBriefs in Climate Studies, DOI 10.1007/978-3-319-41667-0_10

subsequent acceleration has been so great that the quantitative difference has turned into a difference in kind.

All this was facilitated by the trans-nationalization of capital, i.e., by corporations whose shareholders and major investors come from everywhere, whose production facilities are spread all over the world, and whose sense of connection to local environments is extremely tenuous where it exists at all. Real estate speculation, agribusiness, tourism, all these and more are seen only as 'profit centres' through the prism of trans-national capital accumulation. That is the global structure determining the context of the separate phenomena of disaster risk.

This does not mean, however, that there are no elements of disaster risk specific to urbanization and land use as such. On the contrary, and as one example, flooding has increased in urban areas due to inadequate drainage, as has flooding from streams whose catchments are entirely within cities. The underlying cause of this has been the gap between very rapid urban population growth and the limited capacities of urban governments to cope with it. Due to poor urban governance, most cities in low- and middle-income countries absorb this growth by letting informal settlements expand, and these settlements, as we have noted, generally occupy the most hazardous and undesirable land available.

Furthermore, as cities grow, extensive risk expands concentrically in the surrounding territory, following road construction and the growth of satellite urban centres in what was previously rural countryside. This often generates a decline in eco-system services, inadequate water management, land-use changes, and lack of planning for population growth. In addition, the environmental transformation of surrounding rural areas through deforestation, mineral extraction, and the aforementioned road construction, increases the incidence of flash floods and landslides. Meanwhile, towns and cities often displace their environmental risks onto rural hinterlands, burdening these hinterlands with pollution and waste while over-extracting water resources. Rapid, unregulated urbanization thus increases disaster risk in cities and rural areas alike.

The growth of these risks is not linear. Risks increase over time through a concatenation of a large number of individual and collective decisions, often involving land speculation, settlement of some areas by the poor and their eviction from others, mismanagement of environmental resources, and weak local governance.

Weak governance can be seen clearly with regard to land use management, or the lack thereof. Rationally regulating what can be done, on which land and how, requires government structures that are at least relatively free from manipulation by powerful economic interests. These interests tend to look at any barriers to their money-making activities as a mortal threat and react accordingly, even if the activities in question tend to generate disasters. Freedom to use the land they own as they see fit is sacred in their view, and limiting it is a sin against property rights. Any government that hopes to regulate land use for the good of a population at risk must therefore have support from those who are most vulnerable. Again we can see that DRR, if it is to be truly effective, must first draw its strength from communities, particularly those most exposed to hazards.

The problem is, the massive shift of the world's population from the countryside to cities represents a huge business opportunity for several economic sectors, particularly real estate development, construction, and speculative capital. According to projections by the World Economic Forum in 2012, more investment in infrastructure and built environment will be required over the next forty years than has occurred over the past four millennia. Consequently, the real estate development and construction sectors are expected to mushroom over the next decade, while speculative capital will expand from financing their projects.

This enormous business opportunity will, however, present a huge challenge for DRR. Much of the new urbanization will take place in hazard-exposed countries like India and in regions with weak disaster risk management capacities such as sub-Saharan Africa. If the expected investment takes place, as it has in the past, without factoring in risk considerations, then the new wave of urbanization will be accompanied by a new wave of disaster risk that will threaten the resilience and sustainability of countries, cities, and businesses alike.

Conversely, if regulations and incentives are put in place to encourage risk-sensitive investment, the new investment wave may turn out to be an opportunity to further disaster risk reduction. Achieving this, however, will require a massive transfer of resources to finance those incentives, and even more crucially will demand a radical strengthening of governance so that regulations will be enforced.

What is clear is that depending on private investors to choose disaster risk reduction over profits is probably unwise. As we saw earlier, businesses are under systemic structural pressure to grow monetarily if they intend to survive, and this 'law of expansion' may make even risk-aware businesses decide to invest in hazard-exposed areas with comparative advantages such as low labour costs and 'flexible' environmental regulations. Disaster risk reduction will then be traded off against high levels of return on capital because the latter is considered sufficient to offset potential losses, especially if those losses can be shared with the general public. Thus even when they have been assessed, risks are too often ignored in expectation of high short-term profits.

More specifically, the short-term profitability of speculative real estate development does not encourage consideration of disaster risks, which may manifest as losses only after the development has been sold. The original investors rarely take responsibility, or are held accountable, for the disaster risk that may be generated and then sold on. This is further complicated by the fact that risks can rarely be attributed to a single investment decision but are produced by layers of successive investments over decades. Some of this can, of course, be attributed to the deregulation of finance that has taken place increasingly over the past four decades, deregulation that has facilitated so-called 'hot money' that can enter an economy at short notice and leave just as rapidly once its short-term profits have been accumulated.

Urbanization is also an ingredient in the rapid expansion of agribusiness in many parts of the world, particularly in low- and medium-income countries. As we have seen earlier, trade agreements have contributed to pushing many small farmers off their properties and into the cities, with the land 'freed up' then sold to agribusiness interests based primarily in rich countries. What is less widely understood, however, is that above and beyond making farmland available, urbanization gives a boost to the role of agribusiness in mediating food production, distribution, and consumption.

While rural populations migrate to cities and become separated from the land required to grow their own food, the market economy, now dominated by agribusiness, increasingly penetrates food production. Agribusiness investments in processing and distributing have come to dominate the food economy in many countries, to the point where most if not all of a country's population is at the mercy of food price fluctuations that are often unrelated to the actual supplies of food available. And as the market for food in cities and for food exports grows, the expansion of agribusiness in the countryside follows apace.

Since the agricultural practices associated with agribusiness tend to be more disaster-generating than those employed in traditional, small-scale holdings, urbanization not only increases disaster risk in cities but often contributes indirectly to similar risk in the countryside. For instance, intensive water use in drought-prone areas greatly exacerbates the disastrous consequences of drought by draining water reserves and emptying underground water tables, and this ends up reducing the amount of arable land available for growing foodstuffs. Similarly, overuse of chemical fertilizers to diminish the time farmland needs to lie fallow undermines the longer-term sustainability of the arable land that is left, even while increasing short-term profits.

Keeping in mind this symbiotic, disaster-generating relationship between urbanization and agribusiness, we will now turn our attention to the latter and to look at its impact on food security.

Chapter 11
Agribusiness and Food Security

Increased food production worldwide is both necessary and desirable, but the way this is achieved can have deleterious effects on disaster risks, especially insofar as the methods used modify eco-systems. Under many present practices, modifying these systems to increase food and fibre production has decreased their regulatory functions, including those that reduce people's exposure to hazards such as fires and floods. Examples include increased landslide hazard on or near deforested hillsides and more frequent storm surges in areas where mangroves have been destroyed. These changes in ecosystem services quite often benefit specific economic interests, which share few or none of these benefits with the poor rural and urban households that often bear the environmental and economic costs.

This can be seen clearly in the expansion of agribusiness in low- and middle-income countries. Agribusiness investment tends to flow toward producing agricultural products for export rather than toward producing food and fibre aimed at the domestic market, nor are the crops chosen to respond to local cultural and historical preferences. Using farmland that can produce foodstuffs needed domestically for, say, growing flowers for export to Europe and elsewhere in the rich world is a clear example of this, but there are others as well. For instance, producing high priced, relatively rare comestibles is another case of this kind of tailored production, taking land away from growing staples for the local population that can be bought at prices reflecting the local standard of living. Manioc must be sold for much less than orchids or hearts of palm.

Private investment is virtually always aimed at markets where effective demand is greatest and where the prices that can be charged are the highest. In principle, these markets are not found in the producing countries themselves except in tiny wealthy enclaves in urban areas, so a great deal of agribusiness expansion is geared toward markets in rich countries. Even when the production is of fruits and vegetables that are part of the local diet, the price of what is produced is set according

M. Gordy, *Disaster Risk Reduction and the Global System*,
SpringerBriefs in Climate Studies, DOI 10.1007/978-3-319-41667-0_11

to global demand, where the buying power of rich countries and regions is effectively determinant. Unless government price controls are put in place and enforced, most of the local population will have been priced out of the market. With the rapid urbanization we've just been looking at, this price volatility has led to massive hunger even when food supplies are available.

Food price rises have stimulated major new investments in agribusiness and in global food production as a whole. Several factors have driven these rises, some of which have a connection to agribusiness expansion. Rising demand connected with population growth is joined by rapid urbanization due to migration from the countryside. This urbanization, as we have seen, renders more people unable to grow their own food and makes them directly dependent on the food market. As the urban market grows, so do the effects of changing consumption patterns led by a growing urban middle class, engendering shifts in production away from food affordable by everyone. High crude oil prices certainly add to this mix, reflecting in part the use of oil-based fertilizers and oil-dependent mechanization, primarily by agribusiness. In addition, the burgeoning market for biofuels to slake the energy appetites of predominantly rich nations (including the so-called 'emerging economies') competes for arable land that could otherwise be used for growing food, thus cutting into food supplies and raising market prices.

These drivers, often based on genuine supply and demand, are not always the direct, proximate causes of food price volatility. Rather, they are underlying factors that often express themselves through certain catalyzing dynamics, including the concentration of production in a few hazard-exposed regions, declining global food stocks due to disastrous meteorological conditions, and the role of the commodities futures markets.

Agricultural futures markets, for example, like futures markets in most commodities, tend not to depend on the reality of supply and demand but reflect to an increasing extent the speculative bets placed by participants in what are essentially financial casinos. Speculation of this sort results in price volatility that in turn invites more speculation in the form of betting on price trends, i.e., selling short or long and making money on the sheer movement of prices. Another vicious circle.

Pure speculation of this kind has increased over the past few years, due to the deregulation of commodities futures markets in the United States of America and the European Union. This has meant that since 2004, US$173 billion has been used by institutional investors to trade in primary commodities such as food. Commodities futures markets have expanded to absorb this new money, increasing liquidity and leading to huge rises in commodities prices, including for agricultural products. And although commodities prices seemed to have reached a top in 2011, volatility has made it possible for traders to profit from price changes in either direction, as we noted earlier. Even disasters such as droughts or flooding are thus money-making possibilities, a fact that has attracted even more investment into these speculative maneuvers. This has rendered the markets more fragile and unpredictable, making the lives of people who are ultimately dependent on them for food even more uncertain.

Profiting from disasters thus strengthens their effects on the populations at risk. One aspect of futures market deregulation has been to make it possible to securitize commodities and make them more easily tradable and leveraged, thus escalating trading velocity and enhancing the profits made from money movement. This makes prices sensitive to the smallest shock, with even the merely perceived risk of a local crop failure or drought magnified through speculation to the point where global food prices rise dramatically.

Rapid increases in agribusiness investment in low-income countries are exacerbating these tendencies, adding to pressure on available land and potentially adding to disaster risk. Large businesses are buying productive, arable land at low prices, particularly in sub-Saharan Africa, and investing in export-oriented agriculture as noted above. Most of the countries involved have a high proportion of their GDP wrapped up in agriculture and are also subject to high levels of food insecurity. The effects on food prices associated with agribusiness expansion are thus felt disproportionately by the local populations in these countries, rendering them even more food insecure.

Governments are abetting this process by leasing land to agribusiness through various forms of public-private partnerships, using sovereign wealth funds and state-owned enterprises among others. They sell or lease this land to increase their monetary wealth, but in the process they divest themselves of irreplaceable natural capital. Furthermore, because the disaster risks associated with agricultural investments are rarely factored into these decisions, shared social and environmental costs remain hidden.

The agribusiness sector has particularly high social and environmental costs that are viewed as external to investment decision-making and are therefore functionally irrelevant. It is estimated, nonetheless, that the costs externalized by agribusiness currently outweigh the earnings of the sector as a whole. If this was taken seriously and the true costs uncovered, it should, and perhaps would, inspire a reassessment of value creation within the industry from both a business and a societal perspective. However, the same considerations noted earlier that allow many disaster risks to remain hidden may keep these costs invisible as well.

Demand for food, water, and energy is expected to grow by 35, 40, and 50 % respectively over the next decade, according to a National Intelligence Council report published in 2012. Because these resources are interconnected and interdependent, problems related to one resource will affect both supply and demand for the others. High water demand from agribusiness, coupled with declining rainfall in a number of crucial agricultural regions, may well lead to a dramatic depletion of non-renewable water resources. This has already begun to occur. And because agriculture accounts for around 70 % of all water used, including both crop and livestock production and production of animal feed, it is clear that the sector can be a powerful driver of increased disaster risk. Since it is currently dominated by agribusiness, the agricultural sector tends to engage in practices that are inherently unsustainable and disaster-generating.

The impact of these practices over the years is being felt by agriculture worldwide, practices like overusing water resources that are in fact encouraged by government in some countries. In India and Egypt, for example, electricity for pumping groundwater, and the water itself, are free if used for agricultural production. This clearly affects levels of water use and abuse, reducing significantly resilience to drought. As an added attraction, the over-extraction of groundwater is irreversibly undermining water quality, with all the human costs that this implies.

From the point of view of the economic self-interest of a country, under-valuing water price or giving water away for free is clearly counterproductive. The cost of water consumption is thereby never fully included in the cost of the final product or in trade, which means that exporting countries are giving away their precious water resources for free, in effect selling exported products such as flowers or tea at below their true cost of production. It is estimated that in the drought-prone regions of the Sahel and the Horn of Africa alone, between five and fifteen billion cubic metres of water are being exported for free in this way every year.

It is clear from this that because of the export orientation of agribusiness, increases in food production in a country do not necessarily translate into greater food accessibility for the local population. In fact, increased food production by agribusiness may transfer disaster and other environmental risks to communities that are already vulnerable, without doing anything to enhance their food security.

For instance, over the past 40 years in Africa, per capita available farmland has halved and the distribution of what is left has become extremely unequal. Furthermore, communities where agribusiness has implanted itself have lost at least some of their access to productive land and grazing areas, while being deprived of water resources which they had used under customary tenure. Often consisting primarily of small-scale subsistence farmers and pastoralists, these communities are inordinately vulnerable even to small shifts in seasonal precipitation, and without backup water resources they can suffer devastating losses of crops and livestock in times of drought.

In addition, studies show that households in sub-Saharan Africa can only partially satisfy their food requirements with subsistence agricultural production. Most food consumed in poor households is purchased or else comes from food aid that often must be paid for as well. Finding money for these purchases depends on farm households selling what they harvest at the time it is harvested, and what is grown must conform to the demands of the market, not to the food needs of the household.

If a harvest is sold when prices are down, which is often the case due to the greater available supply at harvest time, a price spike later on in the year will cause a household to go hungry because the money they received for their own production will not cover their food needs at current prices. That helps explain what we noted earlier as the persistence of hunger during times of good harvests, and indicates that this is not necessarily a result of food unavailability but rather of inaccessibility due to severely unequal distribution and lack of food price regulation. These problems must be addressed by changing the approach to agricultural governance.

For decades, governments in poor countries have depended on international food aid to deal with food insecurity among smallholder farmers and pastoralists, while simultaneously focusing their own efforts on promoting and expanding export-oriented food production in concert with agribusiness firms. Because food aid is becoming increasingly unsustainable, due to diminishing global surpluses and tightening austerity measures imposed in rich countries, continuing dependence on this aid is institutionalizing food insecurity and its attendant disaster risks.

In addition, present practices in food production and consumption are incredibly profligate. Several studies have shown that roughly one-third of all food produced globally for human consumption is wasted, with between 30 and 50 % of this food never reaching human mouths. Mismanagement of harvesting and storage accounts for a part of this in low-income countries, along with financial and technical limitations. In richer countries, where waste is proportionally much greater, it has to do with consumption habits and market factors. Given the dependence of food insecure households on purchased food, investments in global logistics to reduce food waste, along with efforts at changing consumption patterns, would probably do more to improve global food security than anything else. This would require market interventions that are not on the horizon, however, so for the moment we are constrained to make improvements in the quality and quantity of production rather than in the equity of food distribution.

In this regard, agriculture needs to be regulated so that the present irrationalities are reduced to a minimum. What is needed is a shift in government policy toward helping small farmers participate in a greater proportion of the country's food production. This can be achieved by enabling smallholders to work together to achieve economies of scale that are based on more traditional agricultural practices than those found in agribusiness. This would mean reorienting food production toward domestic needs first of all, and only when food security was firmly anchored turning toward production for export. While that would entail a certain distancing from international trade agreements, it is nonetheless an obvious way to make progress in breaking the chains of rural poverty and a perpetual dependence on 'the kindness of strangers'.

Some of the resources controlled by agribusiness could be re-channeled to small farmers, effectively reorienting the attention of agribusiness itself toward the well-being of the local population. An overall government agricultural policy would be required for this, one where local needs were given priority and where decisions were made impervious to manipulation by any private economic interest.

Given what we know about the relationship between central governments and the decisive economic players, this may seem to be wishful thinking, but generating political pressure through grassroots organizing at the community level might make it possible nonetheless. As in so many of the cases we have been looking at, this is certainly worth a serious try.

There is another important benefit to following this path. Reducing the impact of dangerous agribusiness practices would have a positive effect on limiting environmental degradation. To achieve this reduction, small farmers could work with agribusiness to institute more sustainable and environmentally healthy practices in producing agricultural products for export. A homogeneity of sustainable practices throughout a country's agriculture sector would improve public health and nutrition, while simultaneously lowering the cost of agricultural inputs by moving away from hyper-mechanization and oil-based fertilizers. All of this would be beneficial to the environment as well.

Chapter 12
Environmental Degradation

In previous sections we have touched on the importance of protecting eco-systems to disaster risk reduction. Such protection is part of an overall effort to stop and reverse environmental degradation in general, an effort that begins by recognizing that the natural environment is a social good that cannot be privatized and must be preserved for the sake of human survival. There are no truly 'private' interests with respect to this issue; there is only a social interest in cleaning up after ourselves and social risks in not doing so. Eco-suicide leaves no survivors.

Economic practices based on private capital accumulation cannot take environmental degradation sufficiently into account because these practices will always treat environmental protections as 'trade-offs' against profit making. But there are no trade-offs possible when human survival is at stake, and monetary considerations can never balance the possibility of the extinction of the human race.

One of the barriers to effective management and mitigation of environmental degradation is that its costs are rarely accounted for in public and private investment decisions. One reason for this is that environmental costs are not easily monetized, in part because their effects are cumulative over the long run and as such are not calculable with any great precision at the moment the decisions are being made. Furthermore, as we have seen before, many environmental risks cannot be measured in terms of money because their impacts may engender disasters to future generations and can end up having a scope that is more or less unpredictable.

For instance, polluting a river with mercury, as happened to the Rhine in the 1980s, may affect lives for centuries because of the long half-life of the element's toxicity. However, population movements over that length of time may alter the number of people who will suffer the pollution's effects. Trying to place a monetary value on the lives shattered is therefore a highly speculative exercise, and the same is true for soil poisoning as well as for many other environmental errors.

© The Author(s) 2016
M. Gordy, *Disaster Risk Reduction and the Global System*,
SpringerBriefs in Climate Studies, DOI 10.1007/978-3-319-41667-0_12

If the risk of these disasters cannot be monetized effectively, it will not be included with risks that can be. Since business investment decisions are all about financial return, environmental degradation is necessarily treated as a side issue. Furthermore, inasmuch as governments are caught up in business logic, talk about environmental risks is not considered to be 'evidence based'. Only when these risks turn into actual disasters are they recognized as such, and by then it is too late.

For example, the polluted lagoon next to the Smithfield Ham plant in Mexico engendered the H1N1 influenza virus of 2009. When the virus spread beyond the impoverished local population, government health authorities began to take notice. During the two years prior to that, however, complaints of respiratory illness from that population went unheeded because the risks of living next to a virtual sea of chemically-toxic pig manure could not (or would not) be factored into business or government calculations.

Even when they are recognized, and attempts are made to calculate them in monetary terms, the risks of disaster from environmental degradation are not included in company balance sheets. However, a recent survey of the environmental costs in eleven key industry sectors showed that those costs rose by 50 % between 2002 and 2010 and are doubling every fourteen years. That such costs do not appear in company calculations is an indication that environmental degradation is being externalized by the enterprises most responsible for it.

The general population thus pays the bills, both monetarily and non-monetarily. One important reason for this is the failure of public authorities to manage environmental risks effectively by holding its perpetrators accountable and by regulating business environmental behaviour.

Environmental profligacy generates disaster risks. For example, business investments in areas like bio-fuels, timber, and agribusiness, especially those that involve clearing tropical forests, often increase wild-land fire hazards. These fires result in major depletions of natural capital and in the loss of shared eco-systems we have already mentioned in earlier sections of this document. Agribusiness investments in drought- prone regions may similarly contribute to degrading land quality and to over-exploiting water resources, all of which increase drought risk. Changes in land use, such as rural land being abandoned because of migration to cities or forests being cleared to create rangeland, can increase the long-term risk of catastrophic fires.

Land degradation is associated with intensive agriculture and over-grazing, leading to the breakdown of traditional agro-ecological systems. It can increase the risk of agricultural drought by reducing the moisture-carrying capacity of the soil, while this soil water deficiency can increase the very land degradation that is its cause because of the loss of vegetation that serves as ground cover. There is thus a feedback loop that is difficult to break once it has started, leading to increasing desertification of previously arable land.

What is remarkable is that food production predictions, and economic forecasts in general, do not fully take into account the effects of environmental degradation, painting a rosier picture of future growth in food supplies and other commodities than may turn out to be the case. This is even truer with respect to climate change, which is not yet included in growth projections at all.

The reason for this is that climate change is a surd element, something that, because its implications are not captured by common rationality, calls into question all sorts of environmental projections and will most likely force us to toss many of our time-honoured expectations into the waste bin. And because climate change magnifies most of the disasters confronting the human race, it is a fit subject for discussion in our final section.

Chapter 13
Climate Change as a Magnifier and Meta-disaster

Everything we have looked at so far is magnified by climate change. For one thing, as mentioned earlier, 80 % of disasters are the result of extreme meteorological events, all of which are becoming more frequent and intense as the climate changes, with the effects of the rest exacerbated as these changes accelerate. We also saw that a near-unanimity of climate scientists believes that climate change is the result of human activities, in particular activities associated with the way we produce, distribute, and consume.

For most of us, it is difficult to think about all the various implications of climate change and their dynamic interaction. The problem is so huge and complex that it transcends the common understanding of the word 'problem'. This is because there is no way to 'fix things' in the usual sense, no possible 'solution' or 'programme' that we can turn to while remaining basically as we are. The only way to respond effectively to the species-threatening meta-disaster we are facing is by altering virtually everything about the material structure of the way we live, a concept that defies our notion of rationality because it requires us to call that notion itself into question. But as we have seen, we will have to protect ourselves from 'our way of life' if we are to escape from the disasters it generates and to avoid the extinction of the human race. That entails stretching our minds as best we can so that we find a new concept of rationality that can encompass the enormity of the situation. And we must find this together; there is no individual answer to what is basically a social problem.

While searching for this paradigmatic transformation, we must maintain our morale so that we do not fall into the attractive trap of giving up, of deciding that because we cannot see what to do at the moment, nothing can be done. We must not, as the saying goes, 'cosmologize our own embarrassment'. So in what follows, which is only a preliminary accounting of a very few aspects of the relationship

© The Author(s) 2016
M. Gordy, *Disaster Risk Reduction and the Global System*,
SpringerBriefs in Climate Studies, DOI 10.1007/978-3-319-41667-0_13

between climate change and particular disasters, it is important to remember that the depressing phenomena considered are not the end of the story.

Climate change intensifies the underlying risk drivers of disasters, for example the interaction between disaster risk and poverty. On the one hand it magnifies weather-related and climatic hazards, while on the other it decreases the resilience of many poor households and communities, impeding their ability to absorb the impacts of disaster losses. This loss of resilience results from, among other things, reduced agricultural productivity, increases in disease vectors, and water shortages.

The intensification of these drivers will magnify even further the uneven distribution of disaster risk between rich and poor countries and between the rich and poor within countries as well. Climate change will thus turbo-charge the disaster risk-poverty nexus, drastically increasing the impacts of disasters on the poor.

The dominant global socio-economic structure plays no small role in this. For example, if climate change intensifies the severity of drought in grain-producing regions, as it will come to do increasingly over the next decades, this will feed into speculative increases in food prices, affecting not only people living in the drought-stricken regions but also poor households in other parts of the world that spend a large proportion of their income on food. That is why global markets are called 'global'.

Faced with chronic food insecurity, and finding their resilience undermined by other factors like disease or conflict, poor rural populations will migrate to cities, adding to all the problems of uncontrolled urbanization we've considered earlier. Since one of these problems is greater vulnerability to flooding, the effects of the increased frequency and intensity of floods will join with climate-based drought increases to squeeze the poor coming and going.

One might begin to suspect that the relative indifference towards climate change shown by the governments of rich countries reflects the fact that the wealthier fractions of the population in those countries are able, for the moment at least, to shut out the worst of its effects.

The widening resource gap between rich and poor is socially constructed and is a structural characteristic of the global economic paradigm, as we have seen. This means that the ability of populations to adapt to the changing climate is also a social construct rather than the outcome of changes in the natural environment. Countries that find it most difficult to adapt will be those with fewer resources to allocate to new infrastructure and technologies that might protect them. They will also have more limited social protection and will experience greater food and energy insecurity. Poor countries will therefore be more vulnerable to disasters. In addition, since many poor countries rely on a single economic sector, when disaster strikes they will suffer greater trade limitations than their wealthier, more diversified, counterparts, and this will add to their poverty.

For more than a century, high-income countries have been the biggest contributors to global greenhouse gas emissions by far, but with globalization, many low- and middle-income countries have increased their part as well. Because climate change modifies and magnifies meteorological and hydrological hazards, it is now the ultimate mechanism for global risk transfer and for generating shared risks. It is a prime example of privatizing the benefits of economic investment and transferring the deleterious environmental effects to future generations as well as to populations whose share in those benefits is virtually non-existent. These latter populations, whose responsibility for greenhouse gas emissions is minimal, will find that their greater vulnerability to disasters will impose disproportionate burdens on them in terms of asset loss and increased mortality.

In these conditions, it is necessary for rich countries to help poor ones mitigate climate change, not only by offering material and technical support but more significantly by making radical reductions in their own emissions. Not only are the rich responsible for most of the greenhouse gases produced to date, but only they are in a position to lower global emissions levels to the point where the most lethal aspects of the oncoming meta-disaster can be avoided. This is because in most low-income countries, emissions are already so low that there is little scope for reducing them. Yet it should not be forgotten that in order for reductions in rich countries to be effective, all countries must follow a low-carbon development path so as not to reproduce the conditions that got humanity into this situation in the first place. The problem, therefore, transcends the issue of moral obligation and requires changes from everyone.

Until the past few years, climate change was seen to be something that happened gradually, with gradual responses envisioned as well. But just as a gradual increase in the temperature of H_2O eventually results in a radical change in its form, e.g., from ice to water or from water to steam, so too the build-up of effects from greenhouse gases is issuing in the kind of extreme events the world has recently been witnessing.

This development was prefigured in GAR 2011, when climate change was still being treated in much of the disaster risk sector as something that was happening through incremental changes in average temperature, sea level, and precipitation. That report, while reiterating the perspective prevailing then, nonetheless also added that the task was to reduce and manage the risks associated with more frequent, severe, and unpredictable extreme weather events, *including those for which there may be no historical precedent.* There are indications that these unprecedented events are now squarely on the agenda.

The International Panel on Climate Change (IPCC) reported in 2012 that it is virtually certain there will be substantial, extreme temperature warming by the end of the 21st century, and it is very likely that this will result in sea level rise that will contribute to extremely high coastal water levels. The frequency of heavy precipitation will increase in many parts of the planet, with higher maximum wind speeds in tropical storms likely in certain areas. Glacial retreat and melting permafrost will contribute to the release of methane, a greenhouse gas that is many times more powerful than carbon dioxide, and this will unleash unpredictable and almost unimaginable climate fluctuations.

More recently, in 2013, the IPCC reported on the anthropogenic nature of climate change and suggested ways to mitigate its worst effects, all the while testifying to the acceleration of trends it had previously noted. Now there are even some reputable climate scientists that consider the IPCC's reports to be too conservative, predicting much faster deterioration of climatic conditions that will jeopardize the survival of humanity in the coming decades.

Nevertheless, climate change is still being discounted among investors in agribusiness and related industries, as are the disaster risks associated with it. Forecasts of increased production of crops, livestock, and fisheries products still often assume 'normal' weather conditions as part of a 'plausible' view of the evolution of global agricultural markets over the next decade. Given the massive amount of both scientific and experiential evidence indicating that something fundamental is changing in the Earth's climate, it is difficult to avoid concluding from this that the implications for the way we live are just too disturbing for the dominant economic players and their entourage of advisers to countenance.

Apart from the common psychological defenses against what is often called 'catastrophism', the financialization of trans-national capital has put in place systemic barriers to taking into account the medium- and long-term risks of climate change. For one thing, in recent years the financial markets have gained the ability to respond very quickly to short-term profit-making opportunities, and this focus has made it more difficult for them to pay attention to longer-term liabilities or systemic risks.

This is because the expectation of many financial investors now is that they will only spend a short time with any particular investment, buying and selling their holdings quickly and exposing themselves to risks only in the short run. The rapidity of these financial movements, often speculative in nature, has been facilitated by improvements and innovation in telecommunications and information technologies that have greatly increased the volatility of financial markets, making longer-term reflections increasingly irrelevant to the main business of making money. Taking the financial meltdown of 2007–2008 as an example, this has led to a certain obliviousness or denial that persists with regard to climate change as well. According to the IPCC, investors exhibit a 'lack of confidence in the materiality of climate change' despite amassed evidence to the contrary.

This lack of long-term investor reflection notwithstanding, information on the risks associated with disasters and climate change (which we have seen are inextricably bound together) is readily available. In 2002, for example, a group of financial institutions predicted that economic losses associated with disaster risk and climate change would amount to US$150 billion per year, a figure that was revised upward after Hurricane Katrina hit the US Gulf Coast in 2005. Now there is a newer estimate, calculated in 2007 by the United Nations Environmental Program, that potential losses due to climate change will reach US$1 trillion per year by 2040. This is not a derisory sum.

It would be possible to continue in this vein, but here we should perhaps conclude by reiterating a basic point, namely, that climate change is a meta-phenomenon that expresses itself through a complex variety of effects, each of which is filled with disaster risk. As these risks are realized, they feed back into climate change itself, so that the complex variety of effects is also a complex unity. Addressing most of the specific disasters we have been considering thus requires thinking about climate change, and yet climate change cannot be looked at fruitfully without seeing it in terms of the disasters it magnifies.

More pertinent is the fact that the meta-disaster that is climate change has been created by human beings, specifically human beings in a particular kind of socio-economic configuration that is treated almost as if it is an immutable object rather than an historical expression of ongoing human interaction. This reification of historical social relations is a major obstacle to changing them consciously, and leads to an endemic feeling of resignation in the face of the image of hopelessness thereby engendered. As mentioned above, struggling to free ourselves of that image is a primary task for anyone who would contribute to the survival of the human race.

Meaningful recommendations about what should be done to reduce the risk of this cataclysmic disaster will depend on the particular circumstances in which action will be taken. Yet the possibilities for fruitful interventions will be enhanced if they are undertaken according to principles that reflect what we've been talking about throughout this long discussion. In the concluding remarks that follow, we will touch on some of these principles in the hope that, if they are applied in specific circumstances, they will guide us toward truly effective policies and actions.

Chapter 14
Conclusion

I started out this essay by saying we would look at the first three published GARs from a non-traditional point of view, one that takes disasters to be endogenous to the way we live rather than as exogenous events against which our way of life must be protected. I adopted this perspective because it affords a way to address disaster risk reduction holistically rather than by treating disasters as discrete events that seem just to happen by themselves.

This appeared to be a useful approach, since in reading the first three GARs it became apparent that our socio-economic paradigm is the thread that runs through virtually all aspects of disaster risk, and that its transformation would establish the context for any truly effective efforts to reduce the incidence and intensity of disasters. It also orients our thinking around the combined political and economic aspects of disaster risk, directing our reflections toward specific recommendations that have at least some chance of bearing fruit. If we can identify particular ways in which the dominant socio-economic system generates disasters and erects barriers to reducing the risks and losses associated with them, there is hope that we can discover courses of DRR action that finally will be effective.

Because, looking over the past decade of DRR work, it is clear that whatever has been achieved has done little, if anything, to slow down the overall acceleration of disaster risk, including climate change and the headlong rush to extinction it prefigures. All the well-intentioned international agreements, global conferences, sophisticated white papers, green papers, and action plans have not produced anything like the kind of outcomes needed.

Taking the perspective outlined here is a better way of getting at the underlying factors of disaster risk. By placing the engine of disaster generation in our socio-economic system, we can begin to explore what lies beneath that system's surface. The hope is to uncover patterns of behaviour and structural dynamics, the

© The Author(s) 2016
M. Gordy, *Disaster Risk Reduction and the Global System*,
SpringerBriefs in Climate Studies, DOI 10.1007/978-3-319-41667-0_14

knowledge of which will make possible practical interventions resulting in fundamental change.

Locating disasters elsewhere would involve mystification, leaving us to ascribe their occurrence to immutable laws of nature or, for some, divine intervention. No scope would be allowed for anything other than defensive action and for picking up the pieces after the fact. Prevention and genuine risk reduction would be off the table. On the contrary, if structured human behaviour is responsible for a problem, then it is in principle possible for human beings to solve that problem. We cannot afford to think otherwise.

So the first principle that should guide the choice of DRR actions is that, in an economic system built on the contradictions described above, there are bound to be irresolvable conflicts of interest that cannot be papered over. Disaster risk reduction that takes aim at the economic generators of risk cannot help but confront barriers set up by powerful and implacable economic interests whose aim is to continue with 'business as usual'. These barriers will often be constructed in the name of 'freedom', particularly market freedom, but as the great journalist Felix Greene said long ago, "Freedom in the market means that the big bear is free to eat the little bear, while the little bear is equally free to eat the big bear."

A second principle is that dominant economic power cannot remain so if it is not allied with political power. Weakening that alliance is therefore essential to effecting the fundamental changes needed for genuine DRR. In order to do this we must identify not only the symbiosis between political and economic power, but also the ways in which they are not identical, ways that show that each has a certain relative autonomy. This will help us determine how the space between them can be used to shift political power in a different direction, one that values human beings and their survival above the private accumulation of capital in the hands of a small minority.

The third principle, which is in a sense the progeny of the first two, is that DRR activities have to be based in communities and must provide a vehicle for community pressure. This must be pressure to transform not only the political environment at all levels but also the social relations of production in each country and region. Those relations are currently structurally incapable, as we have indicated earlier, of allowing the welfare of the community as a whole to prevail over the interests of an economic elite and its political retainers.

Communities organized around their own perceived needs, and operating through some form of truly participatory democracy, offer the best way forward, especially if they ally themselves with one another to form a new kind of power bloc that counterbalances centralized and distant political authority. Their DRR actions must be devised not only to address specific disasters but also to target the disaster generators, particularly as these generators reflect the dominant economic model and the contradictions it embodies. This must include, first of all, new ways of speaking about disaster risk and of evaluating disaster losses that do not partake of the conceptual apparatus of business, e.g., universal monetization, the inviolability of property relations, and the immutable goal of GDP growth.

These three principles are not meant to be exhaustive. Many more may be uncovered as our reflections continue. They do, however, provide a starting point that is based on most, if not all, of what has been noted in this document. They are offered as guidance rather than as recommendations, because the latter would necessarily be misleading. Whatever actions are chosen in specific circumstances will obviously have to take account of those circumstances and of their historically and culturally determined conditions. One size definitely does not fit all, and so concrete, practical recommendations must be made in particular socio-historical contexts rather than in general. The people directly involved in the work will have to come up with specific plans.

That does not mean that these principles are useless in a practical sense, but it is evident that they do not allow those persons responsible for DRR, or anyone else for that matter, to abdicate responsibility for making their own judgments. Principles do not apply themselves; their application requires people to make judgments, and there can never be a blueprint for doing this. Yet if we accept them as principles, they can help us avoid repeating the mistakes that have been made up until now, leaving us to make the inevitable new ones that accompany any efforts to do something different. Most important, they can help us learn how to live in a way that will allow our human adventure to continue.

References

UN International Strategy for Disaster Risk Reduction (UNISDR), *Global Assessment Report on Disaster Risk Reduction 2009*; United Nations, Geneva, Switzerland, 2009. http://www.preventionweb.net/english/hyogo/gar/report/index.php?id=9413

UNISDR, *Global Assessment Report on Disaster Risk Reduction 2011;* United Nations, Geneva, Switzerland, 2011. http://www.preventionweb.net/english/hyogo/gar/2011/en/home/download.html

UNISDR. *Global Assessment Report on Disaster Risk Reduction 2013;* United Nations, Geneva, Switzerland, 2013. http://www.preventionweb.net/english/hyogo/gar/2013/en/home/download.html

UNISDR *Global Assessment Report on Disaster Risk Reduction 2015;* United Nations, Geneva, Switzerland, 2015. http://www.preventionweb.net/english/hyogo/gar/2015/en/home/download.html

© The Author(s) 2016
M. Gordy, *Disaster Risk Reduction and the Global System,*
SpringerBriefs in Climate Studies, DOI 10.1007/978-3-319-41667-0

Printed in the United States
By Bookmasters